Reframing
Environmental
Problems

Reframing Environmental Problems:

A Case Study on Wild Pacific Salmon

David Earl Lane

Book cover photo, "Adams River Salmon Run" is copyright © Lee Shoal and made available under a Flickr Creative Commons Attribution License at:
http://www.flickr.com/photos/leeshoal/5293466297/

This book was adapted from:

A DISSERTATION
originally titled

Radically Reframing Environmental Problems:
The *Salmon 2100* Case Study

submitted to and accepted by
OREGON STATE UNIVERSITY

Additional copies of this volume are available for sale online at:
www.BooksaMillion.com
www.BarnesandNoble.com
www.Amazon.com

ISBN-13: 978-1479316021

ISBN-10: 1479316024

DEDICATION

I dedicate this book to my dad, Robert Orrell Lane.

Contents

List of Figures

List of Tables

ACKNOWLEDGMENTS

I would like to thank my committee members: Jonathan King, David Bella, Richard Clinton, Juan Trujillo, and Ron Doel. This case study would not have been possible without the important work of Robert Lackey, Denise Lach, and Sally Duncan, editors of *Salmon 2100*. I give many thanks to the Environmental Sciences Graduate Program at Oregon State University and to the Department of Foreign Languages and Literatures for its continued support via the Graduate Teaching Assistantship in Spanish. I would also like to thank my friends, family, and all my relations. In the publication of this book, I especially appreciate my literary coach David Sanford for the guidance and my brother Aaron Lane for the artwork on the title page. Many thanks to both Aaron Lane and Jennie Cramer for their cover design ideas and support. Additional thanks go to Nancy Raskauskas for editing the manuscript and Justin Katz for tidying up the title page. Most recently, I am filled with gratitude for my new son and teacher, Harley Orion Metolius Lane. Finally, I would like to thank all of the sources I have cited and read.

Preface

As a teacher of Spanish and English, I have experience translating between Spanish and English. However, in this book I am a frame translator. I am translating between different cognitive frames within the same language (in this case, English). I use cognitive linguist George Lakoff's definition of frames throughout the book:

> Frames are mental structures that shape the way we see the world. As a result, they shape the goals we seek, the plans we make, the way we act, and what counts as a good or bad outcome of our actions (Lakoff, 2004, p. xv).

In addition, I have added American sociologist Todd Gitlin's frame definition because he has summarized these frame elements "most eloquently in his widely quoted (e.g., Miller, 1997, p 367; Miller and Riechert, 2001, p. 115) elaboration of the frame concept" (Koenig, 2009): "Frames are principles of selection, emphasis and presentation composed of little tacit theories about what exists, what happens, and what matters" (Gitlin, 1980, p. 6). Communications Professor Michael Maher (who studies framing in the media), states that this definition is one of the best, theoretically speaking, and notes the difficulty of detecting frames:

While it is hard to improve theoretically on this definition, the trouble starts when it comes to the identification and measurement of frames. Precisely because frames consist of tacit rather than overt conjectures, notorious difficulties to empirically identify frames arise (Maher, 2001: p.84).

Some framing researchers suggest that the appropriate metaphor for a cognitive frame is a picture frame (Tankard, 2001, p. 98f; Tankard et al., 1991). Following from this metaphor, frames limit the view and focus the attention and awareness of the reader or listener. Also, the picture frame affects the perspective of the foreground and the background.

While social scientist Thomas König (also spelled Koenig) doubts that any metaphors are suitable for inclusion in sociological theories, the picture frame metaphor is definitely not the metaphor that Erving Goffman chose to use in his seminal essay entitled *Frame Analysis* (Goffman, 1974; König, 2009). In fact, it seems that Goffman (1974) initially avoided the picture frame metaphor and instead used the term "framework" over the first few pages. In particular, Goffman's frames do not *limit*, but rather *enable* the perception of and communication of (social and physical) reality. For Goffman and Gitlin, frames are indispensible for communications; they are the scaffolds for any credible stories (König, 2009).

König (2009) does acknowledge that while viewing frames as picture frames might have its merits, e.g., for use in agenda-setting approaches, he tends to consider them as both consciously adopted but, more frequently, unconsciously used conceptual scaffolds.

In this book I will use the metaphor of the picture frame, because when many environmental scientists frame problems, they are engaging in a kind of agenda setting process that limits and focuses the problem.[1] Think of what we—environmental scientists and others—do when we focus upon a problem to find a solution to the problem in order to not get lost in a torrent of other matters. We focus upon a problem for practical purposes, so that we can identify actions that can be put to use rather than becoming lost in confusion and babble. We frame a problem much as we frame a picture: the frame provides a border that draws attention, directs our focus, to what is enclosed. Note how the words *attention* and *focus* are related to the words *border* and *enclosed*. We frame problems to enable disciplined direction to problem solving efforts so that a particular problem of concern is indeed addressed.

In order to avoid confusion in language in this book, the picture frame metaphor equals *frame*. The scaffold metaphor is *framework*. If the professors, environmental scientists, and policy advocates are framing environmental problems, on one hand, and the students and readers are learning about the problems, on the

---

[1] There is a significant difference between doing environmental science research and making policy recommendations. This book is concerned with both, but only insofar as we are looking at how an environmental problem has been framed. For a distinction on the research and policy making, see Roger Pielke's *The Honest Broker* (Pielke, 2007).

other, then by examining the frames that students and teachers use to address environmental problems we can learn how different frames lead to different views of what are realistic solutions. I chose a case study in order to examine framing and reframing. *Salmon 2100: The Future of Wild Pacific Salmon* was an ideal case because the editors Robert Lackey, Denise Lach, and Sally Duncan clearly framed the problem in Chapter 3, and then they assigned 33 different authors from a variety of disciplines the question "What is it *really* going to take to have wild salmon populations in significant, sustainable numbers through 2100?" (Lackey, Lach, and Duncan, 2006, p. 3). At the time of publication Robert Lackey was a senior fisheries biologist at the U.S. Environmental Protection Agency in Corvallis, Oregon, courtesy professor of fisheries science, and adjunct professor of political science at Oregon State University; Denise Lach was an associate professor of sociology at OSU; and Sally Duncan was a research associate in forest resources at OSU. The premise to the publication is that "it is likely that society will continue to chase the illusion that wild salmon runs can be restored without massive changes in the number, lifestyle, and philosophy of the human occupants of the western United States and Canada" (Lackey et al., 2006, p. 3). According to the editors,

> We know and understand the direct causes of the
> decline of wild salmon numbers. The trajectory
> remains downward. Nothing will change unless we
> address the core policy drivers of this trend: the
> rules of commerce, particularly market
> globalization; the increasing demand for natural

resources, especially high-quality water; the
unmentionable human population growth in the
region; and individual and collective preferences
regarding life style. Do we, as a society, understand
the connections? Can we, and do we want to, turn
the ship around (Lackey et al., 2006, p. 3)?

My role as translator in the *Salmon 2100* case study is first to seek
to understand Lackey et al.'s framing of the problem, second to
seek evidence of any radical reframing of the problem by the
contributing authors via frame analysis methods, and third, to
interpret the different frames. Because reframing a problem often
involves new language that might seem foreign, translation is
required. In addition, because "Every translation is an
interpretation" (Heim, 2009)[2], I will interpret any reframings of
the wild Pacific salmon crisis in *Salmon 2100*. Because authors
who write from a radically reframed perspective have different
frames that may not be well understood, I will attempt to
translate. This is what I term "frame translation." And as with
translation between languages, the translator cannot merely
translate word for word. According to Allison Beeby Lonsdale, who
is a translation professor in the Faculty of Translators and
Interpreters at the Universitat Autònoma de Barcelona, "The main
stages of the translation process are comprehension,

---

[2] Note that this deviates from the way the words *translation* and
interpretation are generally used in translation theory. In most cases,
translation refers to a process involving the written word, and
interpretation is a process involving the spoken word (Beeby Lonsdale,
1996).

deverbalization, and reformulation" (Beeby Lonsdale, 1996, p. 46). Comprehension is a process of analyzing the source language text in order to understand its meaning; deverbalization is a way of then deconstructing the language while keeping the essential meaning in mind; finally reformulation synthesizes that deverbalized meaning into the target language text. Translation involves carefully listening to meaning in both the written and spoken words. Translators and interpreters must be sensitive to misunderstandings, frustrations and people talking past each other because these phenomena may point to conceptual differences that are far deeper than mere differences in word use and sentence structure. Often, body language and intonation patterns may convey meaning that is difficult to translate through mere word selection.

As a frame translator, I deliberately took time to discuss the frames examined in this book with students at Oregon State University. During our discussions, I observed their emphatic gestures that conveyed matters of importance not addressed in their formal education. I discussed the frames contained in this thesis with a group of 3-6 students two hours per week for ten weeks (BA 406). Often the radically new and different frames were more difficult to translate. Linguistically speaking, difficulty in translation between languages depends on the presence or absence of cognates; not just lexical cognates, but structural ones as well. So, the fewer similar words and structures that exist between two different cognitive frames might make frame translation more difficult also.

Like translation between two different languages, this can be a very difficult task. When translating a difficult text from one language to another, the translator may come across difficult words that simply do not translate into the new language. There are at least three reasons for this:

1) There are no words in the new language that convey the same meaning,

2) There are no concepts in the culture of the new language that would even allow for quick comprehension of the difficult words, and

3) There are neither words nor concepts in the 'target language' that adequately communicate the ideas expressed in the 'source language.'

For example, the word *emergence* could be very difficult to understand if it were not already present in a person's cognitive frame. Professor Jeffrey Goldstein in the School of Business at Adelphi University provides a definition of emergence in the journal, *Emergence* (Goldstein 1999). Goldstein defines emergence as: "the arising of novel and coherent structures, patterns, and properties during the process of self-organization in complex systems" (Goldstein, 1999, p. 49).

Peter Corning elaborated Goldstein's definition:

> The common characteristics are: (1) radical novelty
> (features not previously observed in the system);
> (2) coherence or correlation (meaning integrated

wholes that maintain themselves over some period of time); (3) A global or macro "level" (i.e., there is some property of "wholeness"); (4) it is the product of a dynamical process (it evolves); and (5) it is "ostensive" — it can be perceived. For good measure, Goldstein throws in supervenience — downward causation (Corning, 2002, p. 22).

What is Goldstein talking about? To someone who has never conceptually or linguistically comprehended emergence, this definition will contain a lot of misunderstood words like: self-organization, complex systems, radical novelty, wholeness, and supervenience. My role is to help translate these words through frame analysis, examples, visual translation tools (figures), and interpretation. For instance, I could translate the concept of emergence with a simple example from music. Just because I can play good notes does not mean I can make good music. The whole is different than the sum of the parts. Furthermore, the work of individual scientists might be metaphorically seen as well played notes. But the sum of the well played notes does not necessarily equal music well played. Emergence implies that the character of the whole cannot be reduced to the character of the parts. This is an example of frame translation. I translated a potentially difficult concept (emergence) in order to help people understand the implications of matters that are beyond their frames.

While this example may make frame translation appear easy, frame translation, as stated earlier, can be a very difficult task because, as Lakoff writes, "If the facts do not fit the frame, the frame stays and the facts bounce off" (Lakoff, 2004, p. 17). So, how do we prevent new facts from bouncing off? In addition, because frames are often tacit (Gitlin, 1980), they can be difficult to reveal. Remember Maher's admonition, "because frames consist of tacit rather than overt conjectures, notorious difficulties to empirically identify frames arise (Maher, 2001: p.84).

Even though frames are often tacit, the frame in the *Salmon 2100* project is relatively explicit. Lackey et al. framed the problem clearly with four core policy drivers and a required question, "What is it *really* going to take to have wild salmon populations in significant, sustainable numbers through 2100?" (Lackey et al. 2006, p. 3). This clear framing gave focus to the problem solving effort. Because the framing of the problem is clear, the project provides a good opportunity to study the framing and reframing of problems.

In addition to his clarity in the *Salmon 2100* project, Lackey's clarity in the "Challenges to Sustaining Diadromous Fishes through 2100: Lessons Learned from Western North America" (Lackey, 2009), provides opportunities to explore how focused approaches (like *Salmon 2100*), widely presumed to be useful because of their focus on the problem and not on the myriad other problems, might well be 'unrealistic.' Focused approaches might be unrealistic in the environmental sciences, because the environment is so complex and interconnected that to reduce it in a narrow approach in one part of the system, might antagonize another part of the system. Sometimes, the focused approach even makes the problem insoluble. For example, in *Soil Not Oil* the physicist, philosopher, and environmentalist Vandana Shiva "reveals what connects humanity's most urgent crises—food insecurity, peak oil, and climate change—and why any attempt to solve one without addressing the others will get us nowhere" (Shiva, 2008, p. back cover).

In response to Lackey et al.'s clear reframing of the problem, the various authors of the *Salmon 2100* case study used their own cognitive frames as they wrote policy prescriptions. In addition, even as a translator I have my own frames which affect my ability to understand and interpret others. Thus, I must always strive to be vigilant for those facts that are bouncing off of my own frames. One way to be vigilant is to communicate with people coming from other frames, and to ask them what I am missing. I am not an expert in nonlinear dynamics, for example, so I do not have the mathematical background that would allow me to understand the nonlinear frame as a professional mathematician in nonlinear dynamics does. However, with both quantitative and qualitative frame analysis methods in the case study, I can identify when an author is using different words that might indicate a 'nonlinear' frame, and then I can interview the author in order to better my understanding. After repeating this method with various authors and focusing on those who are speaking and/or writing with the most radically different words, I can try to uncover radically different frames. Radically different frames are important to environmental science because the way a problem is framed sets the stage for the kinds of solutions that arise. Radically different frames entail radically different solutions and radically different views of what is realistic.

CHAPTER 1

The Hopeless Dilemma

The impetus for this book came from my experience as an undergraduate and graduate student in the Environmental Sciences Program at Oregon State University. While taking many classes in ecology, human ecology, geology, environmental sociology, and the general sciences, I was taught about the many local, regional, and global environmental problems that exist in our world today. The problems of overpopulation, wars over dwindling natural resources, toxins in the environment, global climate change, and loss of biodiversity were drilled into me. I was often left feeling overwhelmed by the size and complexity of these problems. The professors inculcated a sense of urgency about the current trends affecting our sustainability on the planet. I was left feeling that if one is "realistic" about the problems of overpopulation, severe poverty, loss of biodiversity, energy, food, public health, economic globalization, and toxins in the environment, then the evidence does not warrant much room for optimism. I got severely depressed and hopeless, and I was not alone! The general sentiment of my classmates was the same. The professors and their texts convinced us that there were indeed major environmental problems, and that there were no easy and, at times, no feasible solutions. The trends drained us of hope. We were in a hopeless dilemma; we could be realistic about these problems and end up submerged in gloom and doom, or we could be optimists despite the problems, but with a lingering feeling of being naïve or delusional. So, the hopeless dilemma succinctly stated is realistic gloom and doom (pessimism), on one hand, and unrealistic optimism, on the other. When I learned that the way that the environmental problems are framed affects the solutions, I realized that framing also affects what is realistic, and therefore reframing these problems might lead to a way of transcending the hopeless dilemma. That is why I wrote this book.

In Comes the MAHB

One recent example of the need to explore reframing in environmental and sustainability problems comes from the Millennium Assessment of Human Behavior, also known as the MAHB (pronounced "mob") (Ehrlich, 2009). Paul Ehrlich and other environmental scientists formed the MAHB to deal with the behavioral aspects of environmental problems. They have explicitly stated that environmental scientists should reframe sustainability problems.

In the September 2009 issue of *Science* magazine, the MAHB was featured in an article called "Dealing with Denial" (Ehrlich, 2009, p. 1605). In 2009, the MAHB made essentially four major claims:

1. Environmental scientists know what the environmental problems are, and more natural science will not help.
2. Environmental scientists know what society needs to do.
3. Society stubbornly refuses to change behavior because of denial and apathy.
4. We need to re-frame the problems.

Regarding the first claim that environmental scientists know what the environmental problems are and more natural science will not help, the MAHB reported,

> Millennial assessments of the environmental problems confronting people of all nations have shown that the problems are severe and, in large part, the product of human activities (Mission, n.d., para. 1).

The major environmental problems are:
- Climate change
- decline of food security
- loss of biological diversity

- depletion of water and other vital resources
  with consequent conflict
- use of unsustainable and environmentally
  malign energy technologies,
- deleterious changes in patterns of land use
  and
- toxification of the planet with unregulated
  pollutants that may be dangerous even in
  traces (Mission, n.d., para. 1)

The central problem is clearly not a need for more natural science (although in many areas it would be very helpful) but rather a need for better understanding of human behaviors and how they can be altered to direct humanity toward a sustainable society before it is too late. According to the MAHB, all of these problems "threaten the human future (Mission, n.d., para. 1)."

Regarding claim 2 (Environmental scientists know what society needs to do) Ehrlich wrote,

There is growing consensus among environmental scientists that the scholarly community has adequately detailed how to deal with the major issues of the human predicament caused by our success as a species – climate disruption, loss of biodiversity and ecosystem services, toxification of the planet, the deterioration of the epidemiological environment, the potential impacts of nuclear war, racism, sexism, economic inequity, and on and on (Ehrlich, 2009, para. 6).

Regarding claim 3 (Society stubbornly refuses to change behavior because of denial and apathy) they stated, "Yet society stubbornly refuses to take comprehensive steps to deal with them and their drivers,

- population growth
- overconsumption by the rich, and
- the deployment of environmentally malign technologies (Mission, n.d., para. 1).

The MAHB's "...aim is to penetrate public apathy and denial and prod social scientists to look into the behavioral aspects of Earth's problems" (Ehrlich, 2009, p. 1605).

Regarding claim 4 (we need to re-frame the environmental problems) they invited a wide audience to do so in a global discussion:

> Through a MAHB inaugural global conference, involving scholars, politicians and a broad spectrum of stakeholders, followed by workshops, research activities, and the construction of a human dimensions portal, the MAHB will begin to re-frame people's definitions of, and solutions to, sustainability problems. (Mission, n.d., para. 2).

The MAHB went even further to describe part of what the reframing should do:

> The MAHB would encourage a global discussion about what human goals should be (i.e., "what people are for") and examine how cultural change can be steered toward creation of a sustainable society (Mission, n.d., para. 2).

Ehrlich stated in 2009 that the MAHB's core group at the time, including atmospheric scientist Stephen Schneider and Donald Kennedy, former editor-in-chief of *Science*, was focusing on getting the MAHB's message out. "A 'world megaconference' was planned for 2011" (Ehrlich, 2009, p. 1605). The key point here is that the MAHB was calling for the reframing of environmental problems. A quick look at the MAHB's website (http://mahb.stanford.edu/) demonstrates that the MAHB has indeed been involved in reframing. As of this publication in 2012, the MAHB's mission is: 1) "Foster, fuel and inspire a global dialogue on the interconnectedness of activities causing environmental degradation and social inequity;" and 2) "Create and implement strategies for shifting human cultures and institutions towards sustainable practices and an equitable and satisfying future" (MAHB Mission, 2011). The MAHB even changed their name. They were the Millennium Assessment of Human Behavior, and now they are the Millennium Alliance for Humanity and the Biosphere. The MAHB has called for reframing and is involved in the reframing of environmental problems.

Salmon 2100 and Lessons Learned

Similar to the MAHB's former mission is the *Salmon 2100* Project, because it is an example of a reframed environmental problem. In the project, Lackey et al. reframed the problem of declining wild Pacific salmon in a way that forces scientists, policy analysts, policy advocates, and the rest of society to confront four core policy drivers that had been previously ignored, left out, or denied in salmon restoration strategies.

After the Salmon 2100 Project was completed, Robert Lackey wrote "Challenges to Sustaining Diadromous Fishes through 2100: Lessons Learned from Western North America," (American Fisheries Society Symposium, 2009). Lackey's lessons learned are:

1) The Marketplace is Fundamental

2) Competition for Scarce Resources is Unyielding

3) The Human Population Exerts a Pervasive Influence

4) Individuals Select from Among Desirable Alternatives

5) Policy Domestication is Ubiquitous, and

6) Delusional Reality is Tempting and Widespread

It is an outstanding paper, because Lackey reframes wild salmon problems in a bold and eloquent way that forces environmental scientists to face up to facts (e.g., growing human populations and increasing consumption) that many have set aside to protect their more optimistic assessments. In so doing, Lackey exposes a deeper and more pervasive problem, what I call 'the hopeless dilemma,' which is 'realistic gloom and doom,' on the one hand, and 'unrealistic optimism,' on the other. This 'hopeless dilemma' pervades environmental problems. Thus, Lackey's efforts have uncovered a matter of fundamental importance, far beyond wild salmon. Lackey declares that,

> Outside reviewers (mostly scientists long active in salmon science, management, or policy) of the individual book chapters concluded that the results were realistic in content and conclusion, but at the same time, many of them encouraged us to abandon the blunt realism and forthright honesty in favor of a more encouraging sense of optimism (Lackey, 2009, p. 616).

Similarly, reacting to reviews of articles he had written, Lackey writes,

> ...several reviewers suggested that if my objective in writing was to help save wild salmon (it was not), then the accurate, realistic message would leave proponents dejected. This common sentiment is captured by the following: You have to give those of us trying to restore wild salmon some hope of success (Lackey, 2009, p. 616).

Lackey concludes that,

> To some, my commentary may not be all that
> uplifting. A greater worry to me is that we will
> probably continue to spend billions of dollars in
> quick-fix restoration and management efforts that
> will in many cases be only marginally successful
> (Lackey, 2009, p. 616).

Lackey also concludes with a challenge:

> Any policy or plan targeted to restore any
> diadromous species must incorporate the above
> lessons learned or that plan will fail. It will be
> added to an already long list of prior, noble,
> earnest, and failed management strategies (Lackey,
> 2009, p. 616).

In summary, like the MAHB, the editors of *Salmon 2100*
claim that scientists know the problems (causes of salmon
decline). However, scientists do not have a consensus on what we
need to do, and society is spending billions of dollars on wild
salmon restoration efforts that are not working. Lackey et al. state
"Billions of dollars have been spent, people's lifestyles have been
affected negatively, and commercial activities altered, but still the
prognosis for the long-term future of wild salmon has not
appreciably changed (Lackey et al., 2006, p. 57). Society continues
to proceed with the status quo because of denial and
"domestication" of the salmon policies.

The editors of *Salmon 2100* did reframe the problem by
banishing delusional optimism and baseless pessimism from the

project. They forced all of the authors to confront (and not deny) the four core policy drivers. In this way, Lackey et al. reframed the problem in a way that they considered more "realistic" than delusional optimism and baseless pessimism. Despite the fact that many reviewers perceived the project as pessimistic regarding the future of wild Pacific salmon, Lackey claims that the role of fisheries scientists ought to be policy neutral. Optimism and pessimism are not within the role of fisheries science, only the blunt and realistic facts. It might be the case that many of the reviewers are experiencing the "hopeless dilemma." But if people are getting depressed by the "hopeless dilemma," then that merely becomes a new fact of the world which environmental scientists will have to confront.

In "Challenges to Sustaining Diadromous Fishes through 2100: Lessons Learned from Western North America," (American Fisheries Society Symposium, 2009) (the volume's lessons learned paper), Lackey states that, "The entire premise of the project was to be blunt, direct, and realistic and to avoid both pessimism and optimism," because "Fisheries scientists should be realistic and avoid being either optimistic or pessimistic" (Lackey, 2009, p. 616).

Part of Lackey's concern about delusional optimism stems from the systemic context that puts "pressure on fisheries scientists, managers, and analysts to avoid explicitly conveying unpleasant facts to the public, senior bureaucrats, and elected or appointed officials" (Lackey, 2009, p. 616). Indeed his Lesson #6 is that "Delusional Reality is Tempting and Widespread" (Lackey,

2009, p. 616). Despite the project's banishment of "delusional optimism and baseless pessimism," Lackey states in his lessons learned that, "During the planning and implementation of the *Salmon 2100* Project and the resulting book, the most fascinating aspect was the recurring suggestion, even a plea, to lighten up and be more optimistic and positive in assessing the future of wild salmon" (Lackey, 2009, p. 616).

The Hopeless Dilemma

There are typically two responses to the environmental problems of our time: 1) doom and gloom and 2) optimism. There is a lot of doom and gloom in the environmental literature (for instance, *The Sixth Extinction: Patterns of Life and the Future of Humankind* (Leakey & Lewin, 1995), *The Limits to Growth* (Meadows et al., 1972) *Limits to Growth: The 30-Year Update* (Meadows et al., 2004), *Collapse: How Societies Choose to Fail or Succeed* (Diamond, 2005), and the *Millennium Ecosystem Assessment* (Retrieved December 30, 2009, from http://www.millenniumassessment.org/). The literature warns that the future portends danger for humans and their life support systems. To be optimistic about the present or the future often opens one to being criticized as unrealistic. The optimist is considered naïve. Similarly, when groups of scientists are reporting signs of success in presentations, seminars, and the literature, they are often accused of delusional optimism. They are seen as being in denial or ignorance of some major problems. In the face of this criticism, the optimists are challenged to be 'realistic' by the pessimists; for their belief is that if one is realistic, then one will see that the present and the future are really full of doom and gloom. Thus, if one is being 'realistic,' then one cannot be an optimist. We are doomed. Therefore, we have an apparent choice between realistic gloom and doom and unrealistic optimism. See Figure 1 for an illustration of this phenomenon.

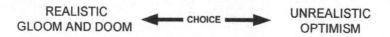

Figure 1. The Hopeless Dilemma: There is an apparent choice between realistic gloom and doom and unrealistic optimism.

But what affects what is realistic? Given Lakoff's definition of frames, the framing and reframing of environmental problems themselves affect what is realistic. Here again is Lakoff's definition of frames, including reframing:

> Frames are mental structures that shape the way we see the world. As a result, they shape the goals we seek, the plans we make, the way we act, and what counts as a good or bad outcome of our actions. In politics our frames shape our social policies and the institutions we form to carry out policies. To change our frames is to change all of this. Reframing is social change (Lakoff, 2004, p. xv).

My thesis is that framing and reframing are relevant in the environmental sciences because they affect what is considered realistic and unrealistic. Therefore, the framing and reframing of problems will have a bearing on the "hopeless dilemma." But first, it is essential to distinguish among several types of framing and reframing. That is the topic of the following chapter.

CHAPTER 2

Framing and Reframing

In this chapter I will describe three types of reframing. The first type is similar to the classic art of rhetorical persuasion. I have termed this type persuasive reframing. The second type expands upon a previous framing of a problem, therefore I have called it expansive reframing. The third kind of reframing changes the reality of the situation entirely, and thus I refer to it as a radical reframing. Now, I will describe these three types of reframing in more detail.

Persuasive Reframing

Lakoff's work in *Don't Think of an Elephant* illustrates what I will term "persuasive reframing." Here, Lakoff is interested in helping "progressive" politicians find the appropriate language to adequately express their worldview and "win the debate." The book focuses on helping the "progressives" frame issues in ways that resonate consistently with the audience and, in so doing, ultimately gain political representation and power. Political pollster Frank Luntz is doing similar work in reframing for the conservatives. His book, *Words That Work: It's Not What You Say, It's What People Hear* (Luntz, 2007) demonstrates some of Luntz's research.

However, within the environmental sciences there has been criticism and controversy surrounding Lakoff's work and the idea of framing in general. Some scientists have rejected reframing as mere political spin. For example, in *Spinning Our Way to Sustainability* (Brulle & Jenkins, 2006), Professor of Sociology and Environmental Science Robert Brulle of Drexel University (Brulle is also a member of the MAHB) and Professor of Sociology Craig Jenkins at Ohio State University argue that "persuasive reframing" is like political "spinning" and that it alone cannot be the proper path to sustainability. Lakoff has responded to such critiques by trying to distinguish reframing from mere spin tactics, but Brulle and Jenkins are not convinced by his argument. They write that although better framing would be useful, alone it can do little. We need to move beyond simplistic analyses and clever spin tactics. What is needed is a new organizational strategy that engages citizens and fosters the development of enlightened self-interest and an awareness of long-term community interests (Brulle & Jenkins, 2006).

In addition, Nisbet and Mooney wrote the provocative article, "Framing Science" in *Science* magazine (Nisbet & Mooney, 2007), to argue that scientists need to learn to reframe problems such as global climate change and evolution in order to be more palatable to the public. They stated that, "scientists should strategically avoid emphasizing the technical details of science when trying to defend it (Nisbet & Mooney, 2007, p. 56). "Framing Science" provoked some flaming letters in response (Holland, 2007; Pleasant, 2007; Quatrano, 2007; Gerst, 2007; and Brewer & Lakoff, 2007). The responses were generally against oversimplifying science through reframing. For example Andrew Pleasant, professor of human ecology at Rutgers (2007) wrote, "Nisbet and Mooney's prescription of framing falls short of a comprehensive diagnosis and treatment plan for what ails science" (Pleasant, 2007, p. 1168). In addition, cognitive linguists Joe Brewer and George Lakoff wrote a response to Nisbet and Mooney, declaring that they misrepresented the science of framing itself. They wrote,

> Framing is not merely linguistic manipulation or a communication strategy, as Nisbet and Mooney suggest. It is the scientific understanding of mental structures built on converging evidence from the many disciplines involved in the cognitive sciences (Brewer & Lakoff, 2007, para. 2).

In summary, Lakoff's reframing strategy in *Don't Think of an Elephant* has ignited criticism and controversy from environmental scientists. There are some who think the ideas of reframing are perfectly suited to the environmental sciences, while others feel that there is something sneaky, tricky, and simplistic about what they see as mere "political spin" tactics. The "persuasive reframing" of environmental problems remains a topic of great debate, and many scientists are not convinced of its appropriateness or utility. So, are there other kinds of reframing that might be more relevant to science? Are frames restricted to "politics" only? Is framing a matter of science? My thesis is that framing does matter in science also. The next types of reframing, "Expansive Reframing" and "Radical Reframing," hold great importance for science. In order to understand how these types of reframing are important to science one must ask the question, how do scientists frame a problem?

Lessons Learned in Science Education

In order to illustrate how scientists frame problems, David Bella at Oregon State University—an emeritus professor of environmental civil engineering and one of the contributing authors in *Salmon 2100*—encourages us to think of a common educational experience in the 'hard' sciences and engineering. A problem is assigned on a test. The problem is framed by the 'given' and 'required.' The 'given' states the facts you (as a student) need to accept to solve the problem. The 'required' states what you are expected to come up with, the result of your solution. Usually, there is a single answer, 'the answer,' the result your solution should obtain. If you do not get 'the answer,' you fail (let us not deal with the troublesome matter of 'partial credit').

We all can draw upon our common experiences as students to uncover important matters that we overlooked as we worked to get "the answer." What should be obvious upon reflecting is this: the frame dictates the answer. Specifically, the answer is pre-defined by the 'given' and 'required.'[3] The challenge for students is to prove to the professor that they can determine 'the answer, that is, 'solve the problem.' But 'the answer' is, in fact, predetermined in the framing of the problem -- by the 'problem setting.'

The frame, however, does not dictate the solution process. There may be a number of different ways to solve the same

---

[3] This is similar to the 'banking model' of education in which teachers 'deposit' information and skills into students. This model was described and rejected by Paulo Freire in chapter two of the *Pedagogy of the Oppressed* (1970).

problem, though all should arrive at the same answer. However, while students may develop solutions that differ, the frame does place constraints on the range of possible solutions. Thus, while the frame does not cut off creative efforts, the frame does give direction to such efforts. Thus, framing a problem and developing a solution are different matters. One can be creative in developing a different solution without changing the frame of a problem.

But in the problem solving experiences of students, they are to accept the frame, not challenge it. Their creativity (originality, knowable innovation) is directed to their solution, not to the 'givens' and 'required' of the problem—its frame. More emphatically, the students are not permitted to change the 'givens' and 'requireds.'

In summary, the common lessons learned by students are these:

1. Accept frames (the 'givens' and 'requireds' of problems assigned).

2. Problem solver's job is to find 'the answer' dictated by the frame.

3. There may be different ways to solve a problem framed by a particular set of 'givens' and 'requireds.'

4. Your creativity and knowledge should be directed to finding a solution that gives 'the answer'; this is known as 'solving the problem.'

Lessons Not Learned

Recall the metaphor of frames as picture frames. Frames limit the view and *focus the attention and awareness* of the reader or listener. As a teacher of Spanish and Environmental Education at OSU, I have experienced the need to reframe an exam when it is beyond the students' capabilities. In engineering Bella experienced this phenomenon also. In interviews with Bella, he insisted that problems are framed for undergraduate students in engineering with 'givens' and 'requireds.'

Shift the perspective now from the students taking a test (solving the problems) to the professor preparing the test (framing the problems). What if none of the students can solve the problem that you the professor gave on a test? When the best students cannot solve the problem, then it would seem that the failure lies not in the students but in your framing of the problem. Bella confirmed this experience in interviews. Professors avoid such an outcome because such a failure points to the professor who framed the problem in a way that it could not be solved, even by the best students. As an example, in a course called 'statics' (an early course in engineering science), a test problem might be "statically indeterminate."

To avoid such troubling outcomes (think how the students would complain!), the professor works out the test problems before including them in a test (professors may also tell Teaching Assistants (T.A.'s) to solve the problems).[4] When professors find

---

[4] Though there are other pedagogical models in science education, Robert DeHaan's article "The Impending Revolution in Undergraduate

that a problem cannot be solved, they reframe the problem. That is, professors change the 'given' and 'required' so that the problem can be solved by the students. However, the professors rarely tell the students that this has been done. Indeed, the professor may be so careful in framing problems (or selecting test proven problems from books and past exams) that the students never encounter a problem that needs to be reframed. And the professors rarely tell the students about the professors' efforts to reframe the problem (so that the problem could be solved by the students). Consequently, the students in problem solving courses fail to see a matter of fundamental importance for problem solving.

The crucial lesson not learned is this: when problems cannot be solved, by competent people, then there are good reasons to reframe the problem (i.e., alter the 'given' and 'required'). Professors know this and act it out (i.e., reframe problems); but they rarely share this with students. And when professors reframe well, the students may never encounter a problem that needs to be reframed. Students learn to accept the 'given' and 'required' of problems. They focus their attention on obtaining 'the answer' dictated by the 'given' and 'required.' And when students do not determine 'the answer,' they, the students, not the frames, are considered to have failed. For the students, the very notion of reframing problems does not come up, even though reframing is a major concern of their professors as they prepare to test the students.

Science Education" (DeHaan, 2005) provides evidence that as of 2005, much of the 'banking model' described by Freire (1970) still exist.

As a teacher who has indeed reframed problems as described above, Bella is forced to confess that the lesson not learned is of paramount importance in our world today. Stated simply, when competent people fail to solve problems (as framed), then there are good reasons to reframe the problem.

Professors and Lessons Not Learned

Let us expand these classroom lessons. Ask a simple question. Should the lessons not learned apply to the professors? That is, in the business of doing their work, are professors missing a matter of importance for the education of their students?

What is taught in university courses is largely the outcome of research conducted by university professors. Members of the science faculty at research universities are busy doing research. Clearly much creative work is done. But, in most cases, research by professors is preceded by a proposal to a funding agency.

Funded proposals frame the research for the research that follows. In science especially, grant administrators in the funding agencies check up on research to make sure it is following the funded frame. Thus, frames do shape research.[5]

But, there is a more fundamental frame, the frame used to

---

[5] Not all research is guided by the frames of funding and this is a complex issue. See James Fairweather and Andrea Beach's "Variations in Faculty Work at Research Universities: Implications for State and Institutional Policy" (Fairweather & Beach, 2002).

write the proposals. Such 'givens' and 'requireds' are embedded in the organizational systems that fund the research. And, over time, successful recipients of funding become the "peer community" that reviews proposals. Collectively they share the frame of thinking that allowed them to be successful proposal writers and recipients of funds. Within this frame, there is much room for creativity, debate, and competition, even for 'new paradigms.' But approaches that fall outside of the common frame are quickly dismissed as 'unrealistic.'[6]

For example, approaches proposed decades ago by economist Herman Daly (Daly, 1978) are very much relevant to the environmental problems of today. But they have been largely set aside as 'unrealistic.' This dismissal has been done with little thought simply because their approaches did not fit the frame. The frame gave direction to thinking, but it was not the subject of thinking itself.

Decades of environmental research have been incorporated into university education. But, if the students who take these courses feel that they are faced with the hopeless dilemma, what should they do? The most common conclusion of research is 'more research is needed.' Another response is 'blame them,' i.e., somebody else –'policy makers,' 'the public,' etc.

---

[6] As noted earlier this book conflates scientific research and policy. Both are regarded insofar as we are talking about the framing of the problem. For example, peer-review, as commonly practiced addresses research proposals, not policy prescriptions. However both research proposals and policy prescriptions can have frames within them.

But when students who are competent and concerned face a hopeless dilemma—a problem that cannot be realistically solved—then the lesson not learned should be taken seriously. As stated earlier, when competent people fail to solve problems (as framed), then there are good reasons to reframe the problems.

But if the frame has been institutionalized within the contexts that define the success and even survival of professors (institutionalized frames), then we are dealing with a form of reframing far more radical than the professors' reframing of test questions for students. For the classroom experience may point to lessons not learned by university professors and grant administrators.

Classroom Lessons Applied to the More Complex Real World

Of course, when dealing with large scale environmental problems in the real world, the connection between frames and answers (policy recommendations, proposals, etc.) is not as precisely defined as in a classroom. There are no single 'correct answers.' Frames, the 'givens' and 'requireds,' are subsumed in professional identities ('this is what we do'), and embedded in organizational structures ('this is what our job is').

Frames, nevertheless, exert their influence. Reflecting upon classroom experiences can be a way to clarify matters of importance that would otherwise be overlooked in the 'real world.' Because such lessons arose from the hard sciences and engineering science, they cannot be dismissed as either soft or impractical. Because they arose from widespread experiences within basic courses, they are both familiar and fundamental. And the stakes involved in the 'real world' are much higher than passing or failing an exam.

In some ways, classroom experiences play a similar role to experimental observations. In both cases we make observations in a simple and controlled environment. Such observations clarify phenomena that would otherwise be hidden and overlooked amid the confusion of the complex and uncontrolled real world.

In the real world, there are better and worse ways to frame and reframe problems. A worse way is to narrow the frame to avoid raising troubling matters. For example, the answers (outcomes, assessments etc.) may thereby become more optimistic than reality warrants—the frame may lead to *unrealistic* optimism. Then, a good case can be made that we must reframe problems to better fit the way the world is rather than the way we would like it to be. 'Expansive reframing' would include these troubling matters as givens. By expanding the givens to include troubling matters that have been avoided, 'expansive reframing' forces problem solvers to address those troubling matters, even if their answers (results) become less optimistic. Of course, there is no guarantee that this will 'solve the reframed problem.'

But if 'expansive reframing' (expanding the 'givens') is done, and competent people still fail to find solutions (as defined by the expanded frame), then what might be done? Some might simply revert to the prior frames -- more optimistic and less realistic.

Others might cynically sink into realistic gloom and doom. But, there is another possible response: radical reframing may be needed to alter our very understanding of what is real (realistic), thereby opening up possibilities that would otherwise be dismissed as 'unrealistic.'

What does radical reframing involve? Here are two statements:

1. Radical Reframing is a fundamental alteration in the ways qualified people 'realistically' address a problem.

2. Radical Reframing alters what informed and competent people accept as a 'realistic' approach to a problem of common concern.

The 'alteration' involves a partial but not complete change that makes a real difference. Thus, radical reframing does not throw out all we know. But it does alter something that we had assumed to be "realistic."

Radical reframing alters what is taken for granted as 'realistic' and 'unrealistic.' It does this in two ways. First, some accepted "facts of the matter" are seen in a radically different way. That is, there is a sharp departure from the usual or accepted way some facts have been framed. Second, the form of the outcome (answer, proposal, and solution) is altered in ways that depart from established (institutionalized, taken-for-granted) practices. There is a sharp departure from the institutionalized ways things are done. That is, established systems are not taken as givens or as the "real world."

To clarify, consider again the 'problem solving' experience of students where problems are framed by the 'given' and 'required.' Radical reframing involves alterations of two kinds:

1. Alteration of a 'given' that changes what is accepted as real ('realistic').

2. Alteration of the 'required' that changes the form of 'the answer' that is sought.

In both cases, the meaning of 'realistic' is changed. In the first case, some data framed as a 'given' are altered. What had been accepted as 'real' is altered in a way that makes a difference. In the second case, the form of the answer required is changed; the nature of what is 'realistically' required—to 'solve the problem'—is changed.

In summary, frames are necessary. They give direction to creative efforts. But when this direction leads to dead ends such as the hopeless dilemma, radical reframing directs creative efforts in different directions. Thus, radical reframing opens up possibilities that would otherwise be dismissed as 'unrealistic' within the earlier frame. However, while the meaning of 'realistic' and 'unrealistic' has shifted, it is in ways that can be justified and defended.

Expected Obstacles and Possible Mistakes

There are obstacles to radical reframing. Resistance should be expected, for when radical reframing alters what is taken to be 'realistic,' it is all too easy to dismiss it with little thought as 'unrealistic.' This can be the 'kiss of death,' for if something is 'unrealistic' it can be trivialized, dismissed, and overlooked. More daunting, radical reframing typically challenges the ways existing organizational systems do things (allocate resources, make assignments, approve proposals, etc.). Radical reframing does not treat these institutionalized systems as 'givens'—as realities. Thus, radical reframing and the answers it leads to will be rejected as 'unrealistic' ('it is not going to happen') because they do not fit— because they challenge—the established systems. Indeed, radical reframers claim that these very systems assume dangerous and unrealistic ways of defining what is 'realistic.'

Having pointed to such expected obstacles, we must also acknowledge mistakes that can be made in radical reframing. I will briefly describe two. The first potential mistake involves a loss of discipline. Without the discipline—based upon a demanding history of careful examination—radical reframing can become little more than the loosening of standards. The outcome can be little more than the proliferation of mere babble and cliché. A second mistake is that—despite new language—the established ways continue. Despite the language of newness, nothing of significance is really changed. For more on this, see "Leading Change: Why Transformation Efforts Fail," by John Kotter from Harvard's Business School (Kotter, 1995) for an acclaimed analysis of implementation failures.

To address such potential mistakes we will look to science and mathematics as a foundation for radical reframing. This may seem strange, even impossible. After all, radical reframing should seek to challenge the devoted work of competent scientists. But this is not as strange as it might seem. The work of competent scientists is largely framed by the organizational systems within which they learn (become educated) and work (apply their education). If we find discoveries in science and mathematics that call into question practices embedded within these systems (the way things are done), then we may indeed find a rigorous basis for claiming that established and institutionalized practices—presumed in the ways things are done—are sometimes, in fact, 'unrealistic.' This would buttress the case for a radical reframing.

We can find a basis for such radical reframing in a paper by Warren Weaver (Weaver, 1948) who was a mathematician and a science manager at the Rockefeller Foundation.[7] As a manager of science, Weaver worked to create new forms of interdisciplinary research in the natural sciences. There is a large body of research that has followed and affirmed Weaver's radical claim. Key words in this radical reframing include 'nonlinear' and 'emergent.' Next, I will further define what was termed 'expansive' and 'radical' reframing in the passages above.

Expansive Reframing

Keeping these lessons learned and not learned from science education in mind, I will now further describe two types of

---

[7] See Robert Kohler's *Partners in Science* (Kohler, 1991) for a major historical work on what Weaver and his colleagues achieved.

reframing that are pertinent to environmental scientists. The first I will term 'expansive reframing.' In contrast to persuasive reframing,' expansive reframing is more than just the use of persuasive language to win debates. This kind of reframing opens a fundamentally new way of seeing what is considered 'realistic' and 'unrealistic' by adding new 'givens' and/or 'requireds' to the problem.

Expansive reframing is when the problem-space is expanded to include more 'facts,' facts that have been denied, ignored, or left out of prior framings. Also, the reframing forces people to consider other problems that interrelate with the original framing of the problem in order to force more of reality onto the situation. The expansive reframer can then call out to a diverse audience to respond to the newly expanded reframing, calling for new creative solutions to the problem that take the expanded aspects into consideration.

One way to illustrate an expansive reframing comes from the lessons learned in science education. In school, teachers assign problems that have a 'given' and a 'required.' As discussed earlier, the students are supposed to look at what they are 'given' and use it to find the 'required' solution. Teachers often do not allow students to change or even question the 'given.' This follows from the 'banking model' of education as footnoted earlier. Again recall DeHaan's article "The Impending Revolution in Undergraduate Science Education" (DeHaan, 2005). In other words,

> There is substantial evidence that scientific
> teaching in the sciences, i.e. teaching that employs

> instructional strategies that encourage
> undergraduates to become actively engaged in their
> own learning, can produce levels of understanding,
> retention and transfer of knowledge that are greater
> than those resulting from traditional lecture/lab
> classes. But widespread acceptance by university
> faculty of new pedagogies and curricular materials
> still lies in the future (DeHaan, 2005, p.253).

In this traditional science education pedagogical framework, students must accept what is given and just proceed to the required solution. The students are then graded based on how well they move from the given to the required solution. An expansive reframing changes what is 'given' and/or 'required,' but the new givens are generally accepted and agreed upon by the new problem solvers in the field even though they (the givens) may have been denied, ignored, and/or left out of prior framings. An example of expansive reframing can be found in Lackey et al.'s *Salmon 2100*. Lackey et al. expanded the givens and requireds in order to be more realistic. However, according to a study done after *Salmon 2100*, all of the project's policy prescriptions were "non-starters" in our current political system. The feasibility study was done by Katharine Whitehead. Whitehead was a Master of Public Policy student at Oregon State University. Denise Lach was her major professor, and Robert Lackey was on her committee. The thesis is entitled, "Assessing the Feasibility of Policy Prescriptions in the Salmon 2100 Project," Whitehead writes,

> In my analysis of the policy prescriptions presented

in the Salmon 2100 project I found that the authors
ascribed a considerable number of burdens to
advantaged populations through the use of
authority tools in their policy prescriptions. This
almost assuredly earns them the dubious
distinction of being labeled immediately as 'non-
starters' in our political system (Whitehead, 2007,
p. 56).

So, what if no competent professionals, as in the case of *Salmon
2100*, can come up with a "realistic" or politically "feasible"
solution to the problem? What do we do when no competent
problem solvers can find a "required" solution to a problem as it
has been framed? The opportunity arises for a radical reframing.
What is radical reframing?

Radical Reframing

What is 'radical reframing' and how does it "really" affect what
is realistic and unrealistic? As defined earlier: Radical reframing
involves frame alterations of two kinds:

1. Alteration of a 'given' that changes what is accepted as real
   ('realistic').

2. Alteration of the 'required' that changes the form of 'the
   answer' that is sought.

In addition, the Oxford Dictionary's definition of radical is as
follows:

radical

> • adjective 1 relating to or affecting the
> fundamental nature of something. 2 advocating
> thorough political or social reform; politically
> extreme. 3 departing from tradition; innovative or
> progressive.

— ORIGIN Latin *radicalis*, from *radix* 'root.'[8]

According to this definition, 'radical' reframing must get to the 'root,' or the fundamental nature of the problem, and depart from tradition in an innovative way that changes what is 'realistic.' When we radically reframe a problem, what we see as real is different. Therefore, what we consider a realistic treatment of the problem is different also.

A consequence of radical reframing is that the radical reframer speaks and writes in ways that may be very difficult for others to understand -- "If the facts do not fit the frame, the frame stays and the facts bounce off" (Lakoff, 2004, p. 17). Thus, the more a person is entrenched in a given frame, the more one views his or her frame as realistic and other frames as unrealistic. When people operating from radically different frames try to communicate, they typically talk right past each other. As a result, they are often frustrated by the other's apparent lack of ability to listen and understand.

---

[8] I am aware that the current meaning of a word is not dependent on its etymology or its meaning in the past, but this definition does offer us some useful meaning.

So, how do we explain such an anomaly of communication when people talk right past each other despite the fact that they are supposedly talking about the same problem? It could be that the two individuals are operating from different frames. However, is it essentially a persuasive reframing, where the language has been reframed simply to gain appeal from the audience and win the debate? Or is it an expansive reframing, where new givens and requireds have been added? Or could it be a radical reframing, where a given and/or a required has been changed or replaced? In order to further distinguish the three types of reframing, the following section provides descriptions and examples of each type.

In contrast to other types of reframing, radical reframing radically affects what is 'real' and therefore it also affects what is 'realistic.' It does this by changing the 'givens' of previous framings in a radical way. When what is considered real is changed, the whole problem may appear new and different. Perhaps more importantly, strategies and policy prescriptions will be different. When what is 'real' (ontological) changes, then ways of knowing (epistemological) typically change. As with expansive reframing, many environmental scientists will agree in theory with some 'radical ideas,' but when it comes to practice, tradition reigns, the radical reframing gets trivialized or viewed as unrealistic, and people go back to 'business as usual.'

Radical Reframing Example: Weaver's "Science and Complexity"

A noteworthy recognition of the need for radical reframing in mainstream science was applied mathematician Warren Weaver's "Science and Complexity," which appeared in the *American Scientist* in 1948. Weaver first showed how science originated by addressing problems of "simplicity" -- problems that only involved one, two, or three variables. Science, especially physical science, made great strides by addressing such problems—e.g., how does one billiard ball affect another when they collide?

Various (not all) scientists at various times moved to the other extreme, looking at problems with millions of variables, as if there were a million billiard balls on the table. Here they used statistical methods to predict how often a ball might hit a given wall or another ball. Weaver called these problems of "disorganized complexity." Problems in between simplicity and disorganized complexity, however, were largely *neglected* as being too difficult. Weaver called these problems of "organized complexity," and argued that we urgently needed methods to address these problems in "the next 50 years":

> These new problems, and the future of the world
> depends on many of them, requires science to make
> a third great advance, an advance that must be even
> greater than the nineteenth-century conquest of
> problems of simplicity or the twentieth-century
> victory over problems of disorganized complexity.
> Science must, over the next 50 years, learn to deal
> with these problems of organized complexity
> (Weaver, 1948, p. 540).

Well, we have passed Weaver's 50 year time limit, and how

far have we come with problems of "organized complexity"? According to Weaver, there were two major wartime advances that could help: 1) computers and 2) "mixed team" approaches of operations analysis. The amazing advances in computation combined with diverse interdisciplinary discourse could lead, as it did during the war, to innovative approaches. As Weaver stated,

> These mixed teams pooled their resources and focused all their different insights on the common problems...and these groups will have members drawn from essentially all fields of science: and that these new ways of working, effectively instrumented by huge computers, will contribute greatly to the advance which the next half century will surely achieve in handling the complex, but essentially organic, problems of the biological and the social sciences (Weaver, 1948, p. 542).

Have we done this?

Weaver nonetheless did not see science as a panacea for all of society's illnesses. He admitted that science is a powerful tool and that it has an impressive record, but "the humble and wise scientist does not expect or hope that science can do everything" (Weaver, 1948, p. 543). In other words, the scientist remembers that science teaches respect for specialized competence, and he/she should not believe that every social, economic, or political emergency would be automatically dissolved if "the scientists" were in control. There are rich and essential parts of human life which are alogical, which are immaterial and non-quantitative in character, and which cannot be seen with a microscope, weighed in a balance, nor caught by the most sensitive microphone. Weaver wrote,

> If science deals with quantitative problems of a purely logical character, if science has no recognition of or concern for value or purpose, how can modern scientific man achieve a balanced good life, in which logic is the companion of beauty, and efficiency is the partner of virtue? In one sense, the answer is very simple: our morals must catch up to our machinery (Weaver, 1948, p. 544).

In order to achieve this 'catching up,' however, knowledge of individual and group behavior must be improved. Communication must be improved between people, and a "revolutionary advance" must be made in our understanding of economic and political factors. A willingness to sacrifice selfish short term interests, either personal or national, in order to bring about long term improvement for all must be developed. According to Weaver, none of these advances can happen unless we understand what science really is with its boundaries included.

Radical Reframing Example: Reinventing the Sacred

A more recent example of radical reframing in biology comes from biologist Stuart Kauffman's *Reinventing the Sacred* (Kauffman, 2008). Here, Kauffman demonstrates how his studies of autocatalytic systems show how emergence is a *real* phenomenon and how the universe exhibits a "ceaseless creativity." This constitutes a radical reframing of traditional science because, in Kauffman's view, the universe is full of self-organizing systems and, hence, is essentially unpredictable. As Kauffman writes,

> Let me pause to explain just how radical this view is. My claim is not simply that we lack sufficient knowledge or wisdom to predict the future evolution of the biosphere, economy, or human culture. It is that these things are inherently beyond prediction (Kauffman, 2008, p. 5).

Kauffman's argument for the inherent unpredictability of reality is indeed radical. What if all of our assumptions that the universe is predictable are false? Kauffman's notions of the unpredictable and self-organizing qualities of the universe make us question just how realistic the trends posited by environmental scientists are. If we assume the universe, evolution, and history unfold in a predictable way, then we will end up in the hopeless dilemma. If, however, the universe is inherently emergent and unpredictable then hope lives.

In addition, Kauffman underscores the importance and relevance of nonlinear dynamics and complexity in modern biology, arguing against "reductionism" as the only path to knowing. He has a lot to say about linear (reductionistic) vs. nonlinear thinking. He claims that the worldview that has dominated our thinking since Newton is reductionistic, which corresponds roughly to American philosopher Stephen Pepper's

"Mechanism" world hypothesis (Pepper, 1961, p. 186; Kauffman, 2008). In his book *World Hypotheses: A Study in Evidence* (Pepper, 1961), he develops a "root metaphor method" and outlines what he considers to be four basically adequate world hypotheses: formism, mechanism, contextualism, and organicsim. The mechanism view of the world leads one to believe that the world can only be understood by breaking it down into its constituent parts.

> The reductionism derived from Galileo and his successors ultimately views reality as particles (or strings) in motion in space. Contemporary physics has two broad theories. The first is Einstein's general relativity, which concerns spacetime and matter and how the two interact such that matter curves space, and curved space "tells" matter how to move. The second is the standard model of particle physics, based on fundamental subatomic particles such as quarks, which are bound to one another by gluons and which make up the complex subatomic particles that then comprise such familiar particles as protons and neutrons, atoms, molecules, and so on. Reductionism in its strongest form holds that all the rest of reality, from organisms to a couple in love on the banks of the Seine, is ultimately nothing but particles or strings in motion. It also holds that, in the end, when the science is done, the explanations for higher-order entities are to be found in lower-order entities. Societies are to be

explained by laws about people, they in turn by laws about organs, then about cells, then about biochemistry, chemistry, and finally physics and particle physics (Kauffman, 2008, p. 3).

Kauffman challenges this predominant worldview, arguing that reductionism alone is not adequate either as a way of doing science or as a way of understanding reality.

Thoughts on Radical Reframing

From the perspective of framing examined in this thesis, the implications of nonlinearity and emergence are indeed radical. Framing—setting boundaries on matters to focus attention to what is enclosed—has been widely institutionalized. Agencies and organizations focus attention on some matters and not others. Within these agencies and organizations, administrative divisions provide further focus. Tasks and assignments provide still more focus to the actual work of virtually everyone. Higher education is similarly structured to focus upon specialized parts and the focused interests of employers and funding agencies.

The very idea that "problems of organized complexity" are extremely important and cannot be reduced to parts poses a radical challenge to the institutionalized focus that is widely taken for granted. Stated simply, the fundamental nature of such problems lies in the character of whole patterns, not in their parts. The institutionalized focus upon parts becomes a form of institutionalized blindness to problems of organized complexity.

Even if solutions to such problems are proposed, they are likely to be dismissed as 'unrealistic' because they do not fit within the focus of any institution. All could say, 'it's not our job,' and consequently proposed solutions would die from lack of attention. This being said, it is not always the case. There are some universities that are already struggling to address larger societal issues at present. For example Arizona State University has adopted a top-down approach to merging departments and colleges to achieve a greater focus on major national and global priorities, including the environment.[9]

If the fundamental insights of nonlinearity and emergence are translated to more common language, they can be seen to pose radical changes to the established ways problems are 'realistically' framed. In fact, the very meaning of 'realistic' and 'unrealistic' is at issue. On the one hand, we find the practical 'reality' that overlooks or dismisses as 'unrealistic' (esoteric) any matter that does not fit the focused frames of established institutions. On the other hand, we find growing evidence that reality itself does not conform to our institutionalized frames. The world is nonlinear, which means that it is unrealistic to presume that parts well done add up to wholes well done.

There are emergent outcomes with serious consequences— "problems of organized complexity" (Weaver, 1948)—that are inherently beyond the borders of institutionalized frames. Radical reframing seeks to address such problems and, in so doing , it

---

[9] For an example of this kind of literature, see "Epistemological Pluralism: Reorganizing Interdisciplinary Research" (Miller et al., 2008).

will—indeed it must—face the rebuttal that it (radical reframing) is "unrealistic" because it does not fit institutionalized frames that focus attention upon countless tasks so that work can be done untroubled by such "esoteric" matters. The radical issue of reframing involves the very meaning of "realistic" and "unrealistic."

A Note on Complexity and Radical Reframing

Graham Harris, adjunct professor at the Centre for Environment at the University of Tasmania and honorary research professor at the Centre for Sustainable Water Management at Lancaster University, wrote *Seeking Sustainability in an Age of Complexity* (Harris, 2007). He describes complexity based on a definition from Gallopin et al., and he links it to Dovers' notion of 'wicked problems.' He writes,

> Gallopin et al. define complexity in terms of the multiplicity of legitimate perspectives, non-linearity, emergence, self-organisation, and multiplicity of scales: in short, Dovers' 'wicked problems.' The factors that particularly impinge on the science of complexity, they claim, are the complexities of physical reality (self-organisation, emergence and uncertainty) combined with the need to consider a plurality of epistemologies and intentionalities (Harris, 2007, p. 171).

According to Harris, we need a new science to transcend and include these perspectives. In 1958 Gregory Bateson wrote of "a science that had, as yet, no satisfactory name". The science that includes organized complexity may have been his referent (Harris, 2007, p. 170). Any radical reframing in this book would capture the ideas of complexity within the "emergent wholes" frames.

Summary

Framing and reframing matter because both affect what is realistic. Because what is realistic affects what is considered gloom and doom and what is considered optimistic, framing and reframing could have an effect on 'the hopeless dilemma.' When scientists expansively reframe, they add 'givens' to the problem that have previously been ignored or denied. But when no competent problem solvers can find a solution to a problem even after expansive reframing, an opportunity arises to do radical reframing. Radical reframing changes a given and/or the required. Through radical reframing new realities emerge and new hopes can transcend the hopeless dilemma. As a frame translator, my task is to translate the differences between Lackey et al.'s expansive reframing and any radical reframing that might exist in *Salmon 2100*. In the case of radical reframing, my charge is to translate (and interpret) potentially strange and foreign words. Words like emergence and nonlinearity will need to be translated through my quantitative and qualitative research methods.

CHAPTER 3

The *Salmon 2100* Case Study

Conceptually, many people may grasp the three kinds of reframing, but often an example is instructive. Therefore, I sought an example or a 'case study' in the environmental sciences that would illustrate both 'expansive' and 'radical' reframing. In particular, I was looking for a case where there was an 'anomaly' in the form of people talking right past each other. While all three types of reframing could be hypothesized as explanations for such an anomaly, I was looking especially for an example of radical reframing because it is the least readily understandable of the three. In particular, I looked for a case where someone used new language to express a new way of seeing the world, of what is 'real' and therefore 'realistic.' I also looked for signs of 'realistic hope,' when 'unrealistic optimism' and 'realistic gloom and doom' seemed to be the only possible reactions to the problem. In *Salmon 2100*, I found all of these characteristics, so I chose to make it my case study.

In the preface to *Salmon 2100* Lackey writes,

> The impetus for the *Salmon 2100* Project can be
> traced to a downtown hotel restaurant table in a
> West Coast city several years ago. Around this table,
> a group of veteran fisheries scientists, policy
> analysts, and salmon bureaucrats mulled over a
> conference they had attended all day (Lackey et al.,
> 2006, p. ix).

Around the table that night, Lackey et al. experienced a different
tone about the future of Wild Pacific Salmon than he experienced
during the day at the conference;

> The tenor of the two discussions was as different as
> night and day. It was almost as if two parallel
> worlds existed, one a fairly positive, optimistic
> perspective about the future of wild salmon, the
> other highly skeptical, pessimistic assessment of
> any of the recovery strategies under consideration
> (Lackey et al., 2006, p. ix).

Given this dichotomy between public and private
assessments of the future of wild salmon, "the overarching
goal of the Salmon 2100 Project is to evaluate critically the
potential options needed to protect and restore wild
salmon runs from mid-British Columbia southward"
(Lackey et al., 2009, p. x). In addition,

> Because the chasm between ecological reality and
> salmon recovery appears to be so immense, both

delusional optimism and baseless pessimism are banished from the project. Instead, we have asked our authors to identify and describe practical policy options that could successfully sustain significant runs of wild salmon if adopted (Lackey et al., 2009, p. x).

In general, the *Salmon 2100* Project *expanded* the frame of the problem to include the realities about the wild Pacific salmon decline by forcing the participants to confront Lackey et al.'s four core policy drivers:

1) Rules of Commerce

2) Increasing Scarcity of Key Natural Resources

3) Regional Human Population Levels, and

4) Individual and Collective Preferences

In order to find authors to confront these core policy drivers Lackey et al.

[...] enlisted more than two dozen salmon scientists, salmon policy analysts, and salmon advocates. They range from hardcore technical scientists to aggressive champions of particular salmon recovery policies, thus representing the spectrum from quasi-institutional to highly individual options (Lackey et al., 2006, p. x).

By expanding previous framings of the wild Pacific salmon crisis, Lackey et al. propose a new 'given' and 'required.' But what if this still does not lead to the 'required' solution? What if the policy prescriptions they receive are seen as "non-starters" (Whitehead, 2007, p. 56) and/or 'unrealistic' given the four core policy drivers? Then, the door opens to a radical reframing.

We know the 'givens' and 'requireds' in *Salmon 2100*, but did any of the contributing authors radically reframe the problem? That is what I asked as I began my quantitative frame analysis. Were the words mentioned earlier, as indicators of radical reframing, present in any of the chapters in *Salmon 2100*? How would one search for keywords that would indicate a radical reframing? The methods section describes my approach to answering these questions.

CHAPTER 4

Frame Analysis Methods

In order to uncover -- to reveal – frames in *Salmon 2100* (Lackey et al, 2006), I did a 'textual/frame analysis' of the primary texts in *Salmon 2100*. First, I ran the primary texts through the Word Cruncher function of ATLAS.ti 6.0 in order to find the frequency of each word used in the different texts. Next, I searched for differences in key word usage within the texts that might indicate a frame difference. The keywords that would indicate a radical reframing come from the writings of Weaver, Kauffman, and others mentioned in the above section on radical reframing (see Figures 4-9 for lists of the keywords used to indicate nonlinear reframing). I then asked the authors if these key words indeed appear to imply a difference in framing. This was a way to validate my initial insights on frame differences and to avoid looking at words that do not indicate a frame difference.

Once keywords were obtained, their inflection forms and synonyms were retrieved from the relevant literature or thesauri or word maps such as WordNet. I then autocoded the keywords (with inflection forms) and synonyms in ATLAS.ti in order to code the alleged frames in the text. I chose ATLAS.ti, because it offers the widest range of autocoding options according to the social scientist Thomas Koenig (2004).

Finally, I attempted to uncover patterns of convergence and divergence in the frames revealed by the dominant frame and any radical reframe. Textual analysis helped me determine the underlying frames of the problem setting and the consequential treatments. This is a replicable method.

After doing a textual analysis of the primary texts, I did another textual analysis of subsequent texts and/or blogs published after the primary texts. I term these 'secondary texts'. I looked to confirm or deny the frames I discovered after the first round of textual analysis on the primary texts. Then I interviewed the authors of the texts in order to validate my findings. I asked the authors if my portrayal of the 'frames' that I uncovered in their respective writings made sense to them.

My method of frame analysis employs both computer-assisted qualitative data frame analysis and other qualitative methods (readings and interviews) in a complementary way. The strength of the computer-assisted qualitative data frame analysis method is that it is fast and automatic. I can autocode the keywords in the entire *Salmon 2100* book in one night. The weakness of the method is that the findings may or may not indicate a frame difference. For example, the same word could be used positively in one case and negatively in another, and yet both cases would be coded with the same code. In addition, word frequency may or may not indicate the importance of the word to the author's frame. So, the qualitative interviews can be done to test whether the frame differences are legitimate or not. Also, one can read through the quotations where the words appear in ATLAS.ti in order to confirm that the keywords are used in ways that indicate the implied frame, i.e., not used negatively or with a different meaning due to context.

In summary, I first personally interviewed both Bella, who did radically reframe the problem, and Lackey to inquire about their respective framings of the problems. I first asked, "How did you frame the problem?" Then I listened and asked follow up questions focused on the frame differences. In addition, I read the texts they wrote in *Salmon 2100* (what I termed 'primary texts') as well as some other texts that they wrote at roughly the period of publication of *Salmon 2100* ("secondary texts"). After reading and hearing that there may indeed be a frame difference in the qualitative data, I did computer-assisted qualitative data frame analyses to continue the search for frame differences with a different method. I then went back to the texts and to the authors to discuss my findings to see if the quantitative data correlated with the qualitative data.

To better communicate the frames and reframes that I uncovered in my frame analysis, I made some maps of my own. Inspired by Wilber's (2001; 2005) "Integral Map," these maps are explained in Chapter 1 and Chapter 6 of this book.

The Robustness Test

Basically there are two kinds of errors that must be avoided if we are to accurately detect a frame difference: 1) where the same word could be used differently in different contexts, e.g., 'whole' systems and 'whole' wheat bread, and 2) the use of different words to indicate a similar frame, e.g., 'nonlinear' and 'chaos.'

In order to detect possible errors of the second kind, a large set of keywords is generated and contrasted with another large set of keywords that may indicate a different frame by using a ratio. The ratio consists of the total number of occurrences of the keywords indicating one frame (Frame A) divided by the total number of occurrences of the keywords indicating a different frame (Frame B).

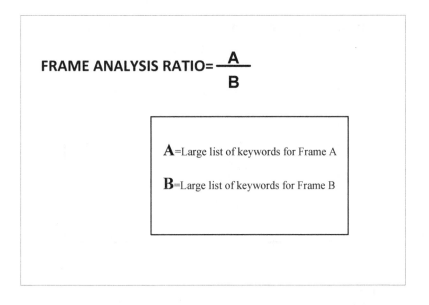

Figure 2. Frame Analysis Ratio

In order to catch the errors of the first kind, I remove words that could easily be used to indicate a frame other than the ones we are looking for. For example, the word 'whole' would be removed from a list of keywords that indicate a frame that acknowledges emergent 'whole' patterns, because whole could be used to mean 'whole' wheat bread, depending on the context. By removing these 'confounding' words we increase the odds of seeing a true frame difference. I call this test the 'robustness test' because if, when we remove words that could easily have different meanings depending on context, there is still a distinct frame difference in the ratio, then we can say that the difference is 'robust.'

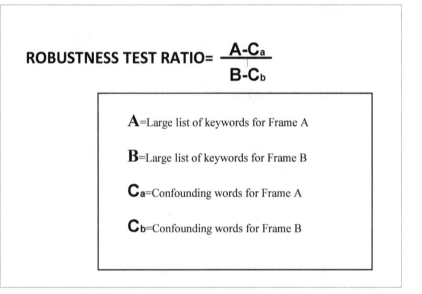

ROBUSTNESS TEST RATIO $= \dfrac{\text{A-C}_a}{\text{B-C}_b}$

A =Large list of keywords for Frame A

B =Large list of keywords for Frame B

Ca =Confounding words for Frame A

Cb =Confounding words for Frame B

Figure 3. Robustness Test

In summary, in order to test the 'robustness' of the frame analysis using Atlas.ti, I start by including as many keywords as possible and then do an autocoding with those words. Next, I remove words that could most easily be confused with other meanings in other contexts. I then do another autocoding (another "run"). If the original findings are not altered much by these tests, then the results can be considered more "robust" than if we had only done an autocoding with the original keywords.

After doing the above described runs in order to seek indications of frame differences between authors of *Salmon 2100* regarding 'nonlinear vs. linear,' and 'we-talk-about-us versus we-talk-about-them,' I realized that there is a third potential error, which I refer to as 'dominant frame language error.' This occurs when an author is speaking from a frame that has been traditionally and institutionally accepted to the extent that it has become 'dominant.' The error lies in the following problem: when people speak from the 'dominant' (common) frame, they do not have to identify the frame or its limitations because it is generally assumed to be accepted by the audience. Whereas, when an author who writes from a new frame, which is different from traditionally and institutionally accepted frames, must identify the new (uncommon) frame and the dominant (common) frame along with their respective limitations. An analogy would be with 'culture.' When a person from the dominant culture enters a room full of people who are also from the dominant culture, they are rarely asked to identify their culture and its strengths and weaknesses. By contrast, if a person from a different culture walks into the room, he or she is often expected to identify where he or she is from, what his or her culture is, and how it is different from the dominant (common) culture.[10]

This analogy should make us realize that, for example, when a person is coming from a linear (common) frame they do not have to say they are coming from a linear frame, and they do not have to express the frame's strengths and weaknesses with respect to any other frame. The frame is just accepted as a 'given,' and they speak outwardly from that frame as a base. In contrast, if someone speaks from a nonlinear frame, they will have to use words to describe their frame and how it differs from the common frame. Herein is a danger in frame analysis. How does one find keyword differences between frames, when the common frame does not even have to express itself? For example, someone from a differing frame might actually use more words to describe the common frame, and how his or her frame differs, than an author who is writing from the common frame itself.

In order to deal with this kind of error, I made a list of keywords that every author used. These would be the keywords of the "dominant" language (common language). I then made a list of the differing frame language (that only some authors used) and made a ratio of the new frame divided by the dominant frame. The extent of deviation from the average then could be used to detect a frame difference from one author to another. In addition, I did runs where I looked at each author's 'different' frame language divided by the average of all of the others (the average minus that particular author). These methods gave me some sense of who had

---

[10] For an example of this phenomenon, see Beverly Tatum's chapter "The Complexity of Identity: "Who Am I?" in *Readings for Diversity and Social Justice* (Adams et al., 2000).

reframed the problem the most in contrast with the others.

The above computer-assisted qualitative data frame analysis methods have several limitations. First, the sample size was too small to do latent class analysis. With a larger sample size, my results could have been analyzed by latent class analysis (a course of future research in frame analysis methods). However, any statistical analysis that tests a null hypothesis is inappropriate for frame analysis for most bodies of text (corpora) because language is never random. Linguist Adam Kilgarriff makes this point clearly in his paper, "Language is Never, Ever, Ever Random" (Kilgarriff, 2005, p. 263):

> Language users never choose words randomly, and language is essentially non-random. Statistical hypothesis testing uses a null hypothesis, which posits randomness. Hence when we look at linguistic phenomena in corpora, the null hypothesis will never be true. Moreover, where there is enough data, we shall (almost) always be able to establish that it is not true (Kilgarriff, 2005, p. 263).

Kilgariff's paper presents experimental evidence of how arbitrary associations between word frequencies and corpora are systematically non-random, and he reviews literature in which hypothesis testing has been used to show how it has often led to unhelpful or misleading results.

The second limitation to my research is that the samples

(authors' texts) were not selected randomly. Lackey et al. chose the authors, seeking diverse viewpoints, but the selection method was not random. If it were random, we could remove selection bias from the study.

CHAPTER 5

Results: Frames, Reframes, and Radical Reframes

Generally, the various runs of the computer-assisted qualitative data frame analysis in both the primary and secondary texts, along with the "robustness test," did indicate that there are some obvious differences in keywords that could imply frame differences between the various authors. There is an indication that Bella has 'radically reframed' the salmon problem and that, compared to the average of the other authors, his language regarding nonlinear, emergent, whole thinking was clearly a strong deviation. However, according to Oregon State University Professors Alan Acock and Juan Trujillo, the small sample size would not yield statistically valid results, so I did not perform a latent class analysis or other such procedures. Even without these methods, the data do show that Bella's chapter sticks out and deviates from 'dominant frame language' in obvious ways with respect to the nonlinear and 'we-talk-about-us versus we-talk-about-them' (see Discussion for this aspect of the radical reframing). The qualitative data from the interviews and readings complemented the quantitative data and allowed me to confirm my hypothesis that there indeed was a radical reframing of the problem in Bella's chapter entitled "Legacy."

Primary Texts

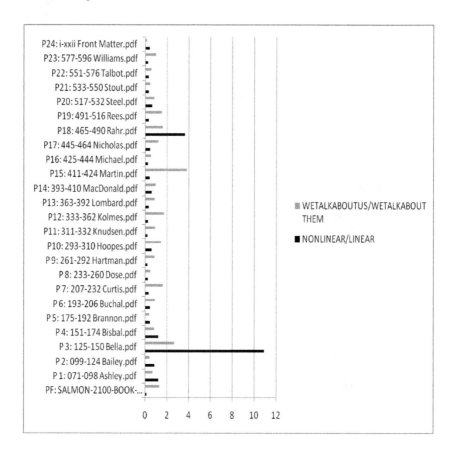

Figure 4. Frame Analysis of Primary Texts (Run 1)

NONLINEAR:=uncertain*|unpredictab*|holistic*|emerg*|chao*|whole*|nonlinear*|complex*|kyros

LINEAR:=certain*|predictab*|analy*|number*|quanti*|trend*|trajector*

WETALKABOUTUS:=we|us|I|me|my

WETALKABOUTTHEM:=they|them|you|he|she|society|societies|

policymakers|fishermen

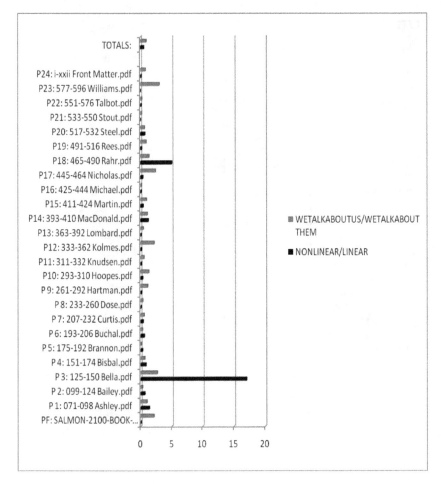

Figure 5. Frame Analysis of Primary Texts (Run 2)

NONLINEAR:=|unpredictable|emerg*|chaos|nonlinear*|
complexity|kyros|holistic

LINEAR:=|predictable|analysis|analytical|numbers|quantit*|
trend*|trajectory|trajectories

WETALKABOUTUS:=us|our

WETALKABOUTTHEM:=them|their

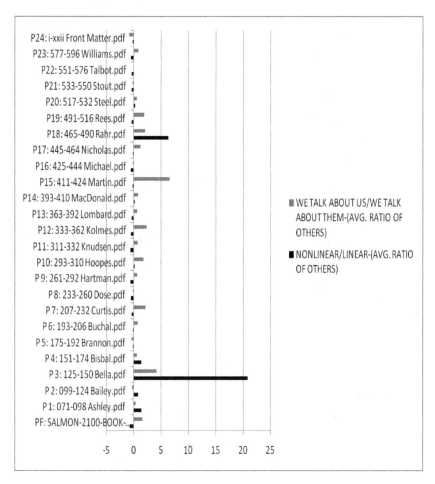

Figure 6. Frame Analysis of Primary Texts (Run 1 Deviations From Others)

NONLINEAR:=uncertain*|unpredictab*|holistic*|emerg*|chao*| whole*|nonlinear*|complex*|kyros

LINEAR:=certain*|predictab*|analy*|number*|quanti*|trend*| trajector*

WETALKABOUTUS:=we|us|I|me|my

WETALKABOUTTHEM:=they|them|you|he|she|society|societies| policymakers|fishermen

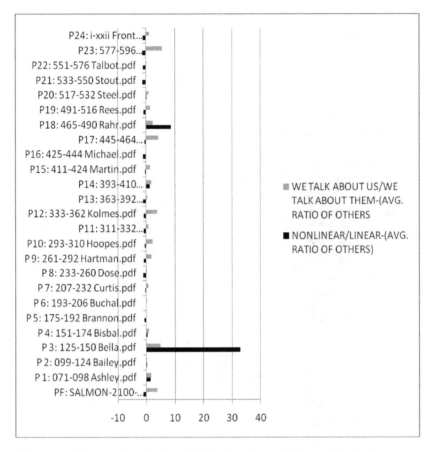

Figure 7. Frame Analysis of Primary Texts (Run 2 Deviations From Others)

NONLINEAR:=|unpredictable|emerg*|chaos|nonlinear*|complexity|kyros|holistic

LINEAR:=|predictable|analysis|analytical|numbers|quantit*|trend*|trajectory|trajectories

WETALKABOUTUS:=us|our

WETALKABOUTTHEM:=them|their

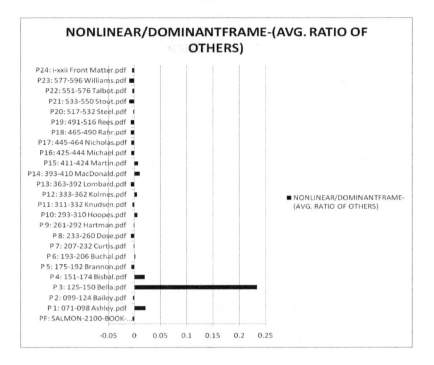

Figure 8. Frame Analysis of Nonlinear Versus Dominant Frame Language. The ratio represents the nonlinear keywords/dominant frame language keywords (Run 3 Deviations of Ratios From Others).

NONLINEAR:=unpredictable|emerg*|chaos|nonlinear*|
complexity theory|kyros|holistic|strange attractor

DOMINANTFRAME:=
conservation|fish|fisheries|future|habitat|harvest|hatchery|
hatcheries|human|I|lake|land|large|life|likely|local|management|
may|more|most|much|must|national|natural|nature|need|new|
number*|ocean|one|Oregon|other|our|over|own|pacific|people|
perhaps|perspective|policy|policies|political|population|
populations|provide|public|quality|recovery|require*|research|
resource|resources|restor*|river*|runs|salmon|science|scientists|
scientific|service|should|significant|societal|society|solution|
solutions|source|sources|spawn|spawning|species|specific|
steelhead|stock|stocks|strategy|strategies|stream|streams|
structure|sustain*|system*|take|their|them|they|things|thinking|
those|through|time|us|value*|water*|we|wild|will|work|world|
would|years|

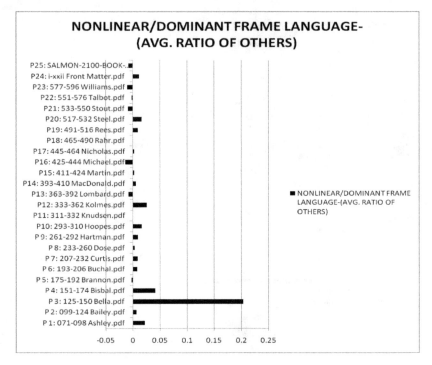

Figure 9. Robustness Test of Nonlinear/Dominant Frame Language

NONLINEAR:=|unpredictab*|emerg*|chao*|nonlinear*|
complex*|

DOMINANTFRAME:=
conservation|fish|fisheries|future|habitat|harvest|hatchery|
hatcheries|human|I|lake|land|large|life|likely|local|management|
may|more|most|much|must|national|natural|nature|need|new|
number*|ocean|one|Oregon|other|our|over|own|pacific|people|
perhaps|perspective|policy|policies|political|population|
populations|provide|public|quality|recovery|require*|research|
resource|resources|restor*|river*|runs|salmon|science|scientists|
scientific|service|should|significant|societal|society|solution|
solutions|source|sources|spawn|spawning|species|specific|
steelhead|stock|stocks|strategy|strategies|stream|streams|
structure|sustain*|system*|take|their|them|they|things|thinking|
those|through|time|us|value*|water*|we|wild|will|work|world|
would|years|

Secondary Texts

I did another frame analysis run on what I am calling the 'Secondary Texts' (other texts that Bella and Lackey wrote at about the same time as *Salmon 2100* as well as texts that were 'lessons learned' from *Salmon 2100*). These runs show that even in the 'Secondary Texts' the authors tended to frame the problems very differently with respect to the 'nonlinear' vs. 'dominant frame language,' and 'we-talk-about-us' vs. 'we-talk-about-them.'

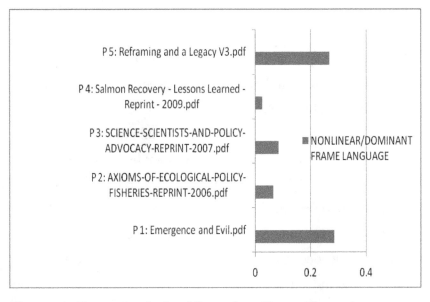

Figure 10. Frame Analysis of Secondary Texts: The ratio represents the nonlinear keywords/dominant frame language keywords (Run 4) (Deviations of Ratios from Others).

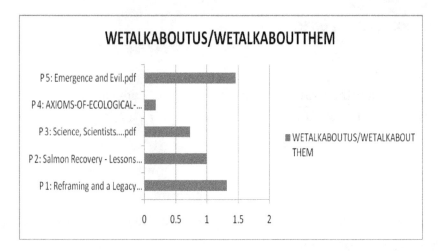

Figure 11. Frame Analysis of Secondary Texts
(Run 4): The ratio represents the 'We-Talk-About-Us/We-Talk-About-Them' keywords (Run 4).

What do these quantitative data tell us? Only one of the authors was different and unusual with respect to the keywords that would indicate a nonlinear frame. This evidence does not prove a radical reframing because it is possible that Bella just gave a different solution to the problem as framed by Lackey et al. without changing a 'given' or 'required.' These data alone do not mean a radical reframing has occurred. Therefore, a qualitative reading of the texts in question, along with interviews of the authors, was needed to determine whether or not a radical reframing had indeed occurred.

Qualitative Data

Cumulative outcomes, irreversible tendencies, nonlinearity, and emergence offer stark challenges to the 'realistic-ness' of core policy driver #4. When Bella first faced the challenge of Lackey et al.'s expansively reframed wild Pacific salmon problem, he thought that, based on the irreversible tendencies in cumulative outcomes, "A Wild Salmon National Park" (Bella, 2006, p. 132) could be a solution. But his proposal for a national park was met with great criticism, and he experienced a crisis. Then, when no solution appeared to follow from the framing that Lackey et al. used, Bella sought to radically reframe the problem. He accepted the first three core policy drivers, but he challenged the fourth. Table 1 summarizes the differences between Lackey et al.'s frame and Bella's reframe.

A Review of the Results

As stated earlier, the qualitative methods entailed a careful reading and interpretation of Lackey et al.'s expansive reframing of the problem (Chapter 3) and Bella's radical reframing of the problem in his "Legacy" chapter. In addition, as noted earlier, I interviewed Lackey, Lach, Duncan, and Bella. In the interviews, I asked for clarification of their respective reframing. Based on the evidence from the quantitative frame analysis and my careful readings of the texts, I asked each author, "How did you frame the problem?" This open-ended question allowed the interviewees to explain to me in their own words how they framed or reframed the problem. They all knew what framing was, so their answers came clearly to me. I then asked follow-up questions in order to further understand the framing of the problem.

After reading the framing Chapter 3 in *Salmon 2100* and interviewing Lackey, Lach, and Duncan, one can find that Lackey et al. did an excellent job framing the salmon "crisis" in *Salmon 2100* (Lackey et al., 2006). It was easy to determine how the problem was framed because Lackey et al. clearly stated their frame. Then, after carefully reading and interviewing Bella, I found that he radically reframed core policy driver #4 and the required question, "What is it *really* going to take to have wild salmon populations in significant, sustainable numbers through 2100?" (Lackey et al. 2006, p. 3).

The *Salmon 2100* Frame

After my readings of the two chapters—"Chapter 3: Wild Salmon in Western North America: Forecasting the Most Likely Status in 2100" and "Legacy"—I found the following. In Chapter 3, Lackey et al. framed the problem in a quantitative way: "What is it *really* going to take to have wild salmon populations in significant, sustainable numbers through 2100" (Lackey et al. 2006, p. 3)? Lackey et al. then stipulate four "Core Policy Drivers" that "must be at the crux of any serious effort to restore wild salmon in California and the Pacific Northwest" (Lackey et al., 2006, p. 61). According to the authors, society *does* control and could *change* these core policy drivers. The drivers are:

1. Rules of Commerce

2. Increasing Scarcity of Key Natural Resources

3. Regional Population Levels

4. Individual and Collective Preferences
(Lackey et al., 2006, pp. 61-68)

Unless changes are made in the current trends of these drivers, the prognosis for the long-term future is that wild salmon populations will continue to decrease. Despite the fact that billions of dollars have been spent, that people's lifestyles have been affected negatively, and that commercial activities have been altered, "the long-term future of wild salmon has not appreciably changed" (Lackey et al., 2006, P. 57).

Lackey sums it up as follows:

> We know and understand the direct causes of the decline of wild salmon numbers. The trajectory remains downward. Nothing will change unless we address the core policy drivers of this trend: the rules of commerce, particularly market globalization; the increasing demand for natural resources, especially high-quality water; the unmentionable human population growth in the region; and individual and collective preferences regarding life style. Do we, as a society, understand the connections? Can we, and do we want to turn the ship around (Lackey et al., 2006, p. 57)?

According to the introduction to *Salmon 2100* entitled "The Challenge of Restoring Wild Salmon," it is likely that society will continue to chase the *illusion* that wild salmon runs can be restored without major changes in the number, lifestyle, and philosophy of the humans of the western United States and Canada (Lackey et al., 2006, p. 3). This *illusion* forms the premise of the entire publication and underlies the challenge they invited the authors to address: "What is it *really* going to take to have wild salmon populations in significant, sustainable numbers through 2100?" (Lackey et al. 2006, p. 3).

Reframing and Legacy: Lessons from *Salmon 2100*

In order to spark some discussion on the Wild Salmon national park proposal, and the radical conceptual challenges behind it, Bella wrote a working paper, "Reframing and Legacy: Lessons from *Salmon 2100*." He distributed the paper to fisheries biologists and "anyone who would take a copy." He presented this material at a "stream team" seminar at OSU. Robert Lackey was in attendance. Drawing on his working paper, Bella summarized the differences in framing as shown in Table 1. All of the statements are quotes from Salmon 2100, except for one which is a shortened paraphrase. Clearly the differences between Lackey et al.'s frame and Bella's reframe are indeed radical.

I interviewed Bella to better understand why he made this radical shift. Specifically, why would he reframe the question shown in the two quotes of Table 1? The answer Bella gave came in the form of a series of unpublished sketches (Figures 12-15). I came to see these sketches as 'visual translation tools.' They are explained and redrawn with a few modifications to show the reasoning behind radical reframing.

Table 1. (p. 72) Lackey et al.'s Frame and Bella's Reframe (From "Reframing and a Legacy")

Lackey et al.'s Frame	Bella's Reframe
The Question:	**Reframed Question:**
"What is it *really* going to take to have wild salmon populations in significant, sustainable numbers through 2100?" (Lackey et al. 2006, p. 3)	"What actions can we take so that the legacy we leave to our children and their children will be honored rather than lamented?" (Bella, 2006, p. 10)
The Failure:	**Reframed Failure:**
Wild salmon populations will continue to decline and many runs will go extinct through this century unless there is a dramatic change in the downward long-term trajectories.	"Wild salmon warn us that our current practices will leave a legacy of loss and lament; this legacy will tell the story of who we really were." (Bella, 2006, p. 10)
What Needs to Change:	**Reframed What Needs to Change:**
"Individual and collective preferences directly determine the future of wild salmon, and substantial and pervasive changes must take place in these preferences if the current long-term, downward trend in wild salmon abundance is to be reversed." (Lackey et al., 2006, p. 133)	"Linear presumptions -- embedded in our institutions, language, and practices -- distort our perceptions, misdirect our actions, and undermine our sense of responsibility; they set the stage for emerging outcomes that are contrary to our highest ideals, our most treasured values, and the faith traditions that many of us hold dear. The decline of wild salmon is a symbol of such emergent outcomes. Radical changes in these presumptions are necessary." (Bella, 2006, p. 10)

Bella's Framing Objections

Confronting Lackey et al.'s framing of the problem, Bella describes two objections to the framing and a prescription:

1) The focus on measurable numbers of salmon will lead to a "technological fix" via hatcheries.

2) Outcomes can emerge that no one prefers, and we have to deal with these.

Regarding objection number one, Bella suggests that framing the problem in a quantitative way, as Lackey et al. do when they ask "What is it *really* going to take to have wild salmon populations in significant, sustainable numbers through 2100" (Lackey et al. 2006, p. 3)?, will lead to the motivation to redefine wild salmon to include hatchery fish and their offspring. Thus, we can "solve" the problem through a "technological fix". Bella states,

> When the strategic standards for success are defined in measurable outcomes, the stage is set for the technological fix; technological solutions such as hatcheries are all but assured (Bella, 2006, p. 139).

Bella makes this counterclaim because he asserts that technological success arises over time by:

- Developing, designing, constructing, and operating some device or structure to meet some measurable performance criteria

- Learning from past failures to meet the criteria and modify actions; and
- Redefining the objectives and performance criteria to fit the fixes (Bella, 2006, p. 139).

If this sequence of technological success were to be applied to the sustainable salmon question, then it seems like the framing of the question as "what is it going to take to have wild salmon populations in significant, sustainable numbers through 2100?" is the first step because it establishes the measurable performance criteria. We could be in the process of the second step right now learning from past salmon fisheries failures around the world and in the Pacific Northwest in order to meet the performance criteria and modify actions. The third step would be to redefine the objectives and performance criteria to fit the fixes. Bella argues that with the first two steps generally comes the third, and, so we may end up redefining salmon in order to fit the likely technological fixes. In this case, it would mean that we could move from defining wild salmon as "those produced by natural spawning in natural or minimally altered fish habitats," to "salmon not harvested from fish farms," which would include hatchery fish.

> The mere promise of technological success often serves as justification to reduce or set aside those actions (protecting natural and healthy ecosystems) that limit or restrict technological development. This chapter asks, 'does the cumulative outcome of such technological development result in a world, a legacy, that we wish to leave for our children's

children?' The chapter answers no, [*sic*] and proposes an alternative. Measurements can, of course, serve tactical or specialized tasks, but tactical tasks must serve strategic or comprehensive purpose [*sic*] and strategic purpose must not be defined only by what we can clearly measure (Bella, 2006, p. 139).

In his unpublished working paper, "Reframing and Legacy," Bella points out that Lackey had already written a paper that gave definitions of "wild salmon," two of which would include hatchery raised salmon—though this has not been accepted yet. If wild salmon are redefined to include hatchery fish, then the "technological fix" could be the result.

Figure 12 illustrates the technological fix that Bella believes will follow from framing the wild salmon crisis in quantitative measurable outcomes. When reading any of Bella's sketches in this book, one should say 'therefore' on an arrow moving forward, and 'because' when moving backward.

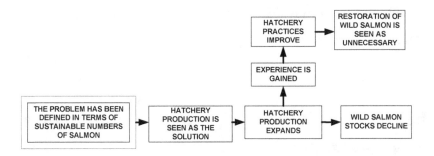

Figure 12. The Technological Fix

Next Bella introduced Figure 13. This figure demonstrates how the protection of ecosystems has been justified in terms of salmon protection. Consequently, if wild salmon decline, then there is danger that the protection of ecosystems is more difficult to justify.

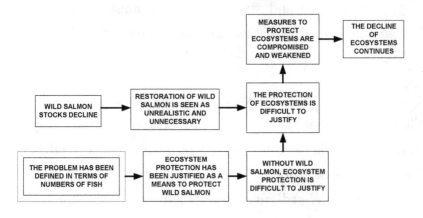

Figure 13. Consequences of Actions on Ecosystems

In addition to the previous two figures, which reveal the potential for "the technological fix" (Figure 12), and the consequences for ecosystem protection if wild salmon stocks decline (Figure 13), Bella added Figure 14 which illustrates the effects of the "irreversibility principle."

The Irreversibility Principle and the Cumulative Outcomes Model

In order to illustrate objection number two from above: "outcomes can emerge that no one prefers, and we have to deal with these," Bella (2006) describes what he calls a cumulative outcomes model. This model depends on the irreversibility principle. The idea is that over time the world will become full of the consequences of the relatively irreversible decisions, even if we only choose them fifty percent of the time.

For example, to understand how unintended outcomes occur as emergent properties within ecological-social-economic systems, Bella asks us to imagine spending several days exclusively devoted to the following: (A) Watching television, playing video games, driving in traffic from mall to mall, and exposing ourselves to advertising of all kinds; or (B) quiet reflection near a stream in a forest away from the busyness of our modern commercial and technological world.

Now, we are to ask ourselves, "What kind of world do we wish to leave for our children?" Does your value system (preferences) cause you to say, "We need to devote effort, time, and resources to producing more of (A) and less of (B) for our children?" Do we really want a world for our children that is even more dominated by (A) at the sacrifice of (B)? Is such a world—with (A) promoted, developed, and hyped, and (B) diminished, fragmented, and eventually eliminated—so important to us that it is worth the large amounts of time, talent, and resources needed to bring it about? Bella has not found a single person who will answer yes. He goes on to say that it is hard to find anyone seriously advocating movements to "Pave the earth," "Kill the salmon," "Celebrate traffic congestion," or "More TV for kids." And yet many people from various faiths and walks of life believe this shift from (B) to

(A) is occurring: (A) is being promoted, developed, and hyped, while (B) is being reduced, fragmented, and eliminated (Bella, 2006, pp. 129-130).

The cumulative outcomes model offers a way of seeing the cumulative outcomes of many decisions involving many different people acting at different times and under different conditions. And, as in the real world, everyone knows relatively few of the others, and the contexts of the decisions vary. It is a dynamic world in which the actions of the people and the consequences of their decisions adapt to changing political climates, new information, shifting economic conditions, changing participants, and different recommendations from experts and professionals (Bella, 2006).

Bella asks us to imagine the whole decision-making process as a game involving flipping coins: Each flip of a coin represents a decision that is made. Whenever the coin comes up heads, the decision involves technological development of some natural resource; e.g., constructing a highway, mining a deposit, or harvesting a stand of timber. When the coin comes up tails, the decision involves the protection of some natural system from technological development; for example, "tails up" outcomes imply decisions to protect old-growth forest or prevent development along a stretch of river. Let us assume that each decision involves equal value placed upon technological development and protecting the natural world, including wild salmon and old-growth forests. Thus, wild salmon and old-growth forests are valued equally with shopping malls, commercial television, and video games, for example. We create such a balance by assuming an unbiased flipping of the coins, which give us equal probability of heads or tails, i.e., a "50-50 chance."

Rather than focusing on individual coin tosses at particular times, however, Bella asks us to consider the cumulative outcomes of all coin tosses over a long span of time. In order to make our imaginary world of tossing coins more realistic, we add a rule: coins that come up heads stay heads; in other words, when a flipped coin lands on heads (resource is developed), it is no longer flipped—the resource stays developed. Whereas when a coin comes up tails (resource is preserved), it can be flipped again at some later time. Now, imagine the cumulative outcome of continued coin tosses. A trend emerges from the whole that cannot be seen by examining individual coin tosses; this trend is the growing dominance of all decisions based on the "heads up" position. Through many unbiased coin tosses, we approach a world where the "heads up" development decisions, which are relatively non-reversible, gradually but steadily increase in number. That is, we approach a world of pervasive technological development where wild salmon, for example, have no place (outcome A), not a world where wild salmon can survive and thrive (outcome B).

Traditional linear thinking assumes that if each individual decision is informed by assessments for each outcome, then the cumulative outcome of all decisions will reflect the values (preferences) of the society. Our cumulative outcomes model suggests something very different, an irreversibility principle, which can be stated:

> The cumulative outcome of many decisions within a
> dynamic system will be dominated by the most
> irreversible tendencies within human actions [*sic*]
> regardless of the values people hold (Bella, 2006, p.
> 132).

Within the context of human activity, each decision (flip of the
coin) may value development and protection equally with a "50-50
chance," and yet, the cumulative outcome of many decisions over
time favors the more irreversible outcomes (i.e., development).

Bella reminds us that this model, like all models, is a
simplification. A model has worth, however, when it allows us to
discern some matters of importance in complex systems, i.e., when
it allows us to see a pattern, a trend, or a tendency that is not
otherwise obvious. Bella asserts that many models of management
fail to address the aforementioned cumulative outcomes and
impacts. For example, the market model does not adequately
address cumulative outcomes emerging from many decisions, and
market ideologues have a bad habit of denying any problem that is
not well accommodated by their model (a fundamental and critical
mistake with any model). Economic models tend to discount
future outcomes. Scientific models tend to define problems in
ways that set the stage for a scientifically derived technological fix.
Common models of human behavior lead one to say, "Outcomes
reflect values; to change outcomes, one must first change values"
(Bella, 2006, p. 132). The flipping coin/cumulative outcomes
model helps us to perceive the irreversibility principle, from which

a practical recommendation emerges that could make a meaningful difference for the world we leave to our children's children: our legacy (Bella, 2006). Based on the irreversibility principle, as illustrated by the cumulative outcomes model, Bella prescribed a wild salmon national park in order to preserve habitat and prevent the irreversible tendencies from taking over completely. Figure 14 illustrates the effects of the irreversibility principle on wild Pacific salmon.

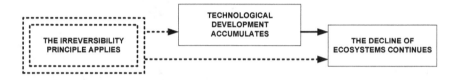

Figure 14. The Irreversibility Principle (Illustrated by the Flipping Coin Model)

The System that Emerges

When the technological fix (Figure 12) is combined with its consequences on ecosystems (Figure 13), and the irreversibility principle (Figure 14), a system emerges from the framing of the problem that is a combination of Figures 12, 13, and 14. Bella combined these Figures to produce Figure 15

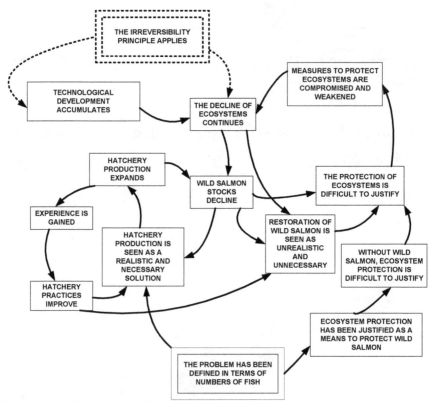

Figure 15. The System that Emerges

The system that emerges from Lackey et al.'s framing of the problem and "The Irreversibility Principle" (Figure 14) identified by Bella

The system that emerges in Figure 15 follows from framing the problem in terms of numbers of fish(see box at bottom of figure with double lines), and the "irreversibility principle" (in dotted box at the top of the figure). This system forced Bella to reframe the question in Table 1 from "What is it *really* going to take to have wild salmon populations in significant, sustainable numbers through 2100?" (Lackey et al. 2006, p. 3) to "What actions can we take so that the legacy we leave to our children and their children will be honored rather than lamented?" (Bella, 2006, p.10). See Figure 16 which includes Figure 15 plus the question, 'Will this lead to a legacy we wish to leave?'

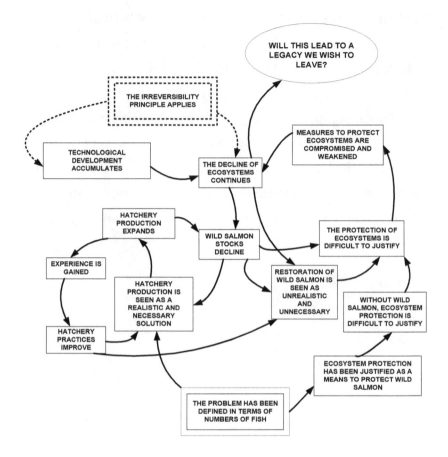

Figure 16. Is the Emergent System a Legacy We Wish to Leave?

In order to partially answer this question, Bella proposed a wild salmon national park.

Bella's Wild Salmon National Park Proposal

The proposed wild salmon national park provides a response to the problems illustrated in Figure 16. The park would provide a relatively irreversible protection to the wild Pacific salmon ecosystem. In addition, the establishment of such a park would be a hopeful action that is not merely a technological fix via hatcheries or fish farm production. In contrast to the technological fix, the proposed park would provide a lasting legacy that includes ecosystems.

Unlike traditional national parks in the United States, the wild salmon national park would be distributed over the Pacific Northwest.[11] It would not be one isolated plot of land. The distributed nature of this park does not fit the established and institutionalized frames of public agencies (local to national).[12]

Crisis and a Radical Reframing

After writing his initial wild salmon national park proposal, however, Bella was heavily critiqued, as noted earlier, and experienced a crisis. This crisis led him to radically reframe the problem and to assess the obstacles to such a reframing.

In an interview, Bella stated that when he wrote up the two objections and the policy prescription of a wild salmon national park, he thought that he was done. Then, upon review and dismissal from a number of critics, he was forced to reframe the problem altogether and undertake what I have described as a

---

[11] To better understand the current national park system, read *The National Parks: America's Best Idea* (Duncan & Burns, 2009).

[12] For a reference on how such a distributed solution could utilize a whole systems approach, see *Natural Capitalism* (Hawken, Lovins, & Lovins, 2000).

radical reframing. For example, he received a scathing review from one environmental scientist that insisted that people's values would never support the wild salmon national park idea. At that point Bella experienced his crisis and felt that he was forced to radically reframe the problem. The reviewer wrote,

> It seems clear to me that the cultural values of this country were reflected in . . .the fact that SUVs substantially dominate the car market, and that the average person watches five hours of TV a day. I disagree that this country cares enough about our children (let alone our obligations to other species) to even pay for a small park, let alone something as grandiose as he's imagining.

This kind of critical review led Bella to reconsider the role of values in determining behaviors. According to Bella, there are at least three major obstacles to the wild salmon national park:

1) Values Determine Behaviors
2) Higher Education
3) "It Won't Make a Bit of Difference"

Obstacle 1: Values Determine Behaviors

According to Bella, the first obstacle involves a common view often strongly stated: before there can be any real change in the current trends of environmental loss (degradation), there must be a fundamental change in society's values. From this point of view, the proposal for a wild salmon national park is rejected almost automatically and without hesitation.

Bella states that such views generally have three parts. First, there is a clear recognition that many destructive behaviors are occurring in our society and that the cumulative outcomes of such behaviors are harmful and potentially catastrophic (he agrees). Second, there is the presumption that these behaviors are determined by society's values (he disagrees). Third, it logically follows that behaviors will not change until society's values change (having rejected the second, he rejects the third).

The second part asserts a values-determine-behavior model that Bella critiques. Unfortunately, many people often fail to acknowledge this as a model; it is asserted as an obvious fact with little or no thought given to alternative models that offer different ways to explain human behavior. From a linear perspective, it makes a lot of sense. Human behaviors are reduced to the values held by the parts, or the individuals, so there is no need to study whole patterns of self-reinforcing behaviors. Then it becomes easy to blame others by saying, "Nothing can be done until they change their values."

An alternative, nonlinear model is that contexts shape behaviors, often in ways that conflict with values (Bella et al. 2003). Contexts emerge as self-reinforcing behavioral patterns that shape the behaviors of those within them, often including scientists, engineers, and highly competent experts in any given field. Horrible outcomes can emerge not just from horrible people but also from competent people, much like ourselves (Milgram, 1974). In order to explain how this can happen, Bella wrote "Emergence and evil" (Bella, 2006). In "Emergence and evil" Bella "demonstrates that emergence, as a disciplined way of thinking, can expand our understanding of evil and responsibility in ways that are relevant and critically important" (Bella, 2006, p. 102). Bella draws from Kurt Richardson's insights (Richardson is the Associate Director of the Institute for the Study of Coherence and Emergence). "An important property of emergent wholes is that they cannot be reduced to their parts... wholes are qualitatively different from their parts... they require a different language to discuss them" (Richardson, 2004: 76-77). In so doing, Bella shows how evil can emerge as a systemic behavioral context in which any one of us could get involved.

Obstacle 2: Higher Education

Firstly, Bella claims that linear presumptions are embedded within the common practices of professors and the habits that students acquire while gaining their education.

Secondly, Bella claims that linear presumptions are built into administrative structures that largely parallel the academic disciplines and reinforce concerns for parts rather than wholes. In addition, faculties are under increasing pressure to gain outside funding, and faculties pursue such funding as a way to protect their own specialized fields. Some limited interdisciplinary work has been done with outside funding, but the net effect has been to direct attention to new and fundable specializations that often fail to accommodate larger nonlinear wholes. Given these administrative structures, pressures to gain outside funding, productivity demands, and the drive to succeed within specialized disciplines, linear presumptions prevail because they make matters more manageable for everyone.

Thirdly, universities have taken on a corporate mentality,

justifying their existence on the basis of economic development. The notion of preparing "an alert and knowledgeable citizenry" (Eisenhower, 1961) for our world of technological development, in Bella's opinion, has been all but given up.

Obstacle 3:
"It Will Not Make a Bit of Difference"

Bella identified the statement "It will not make a bit of difference," as another major obstacle to the wild salmon national park. He suggests that the political climate of today is all about economic development—i.e., more stuff—not quiet refuges that actually let something be free and wild. According to Bella, our world is focused on short-term profit and political gain, not the long-term legacy for our children's children. We often see ourselves as producers and consumers, not remnants of an ancient nonmaterialist tradition. And even if we decide to do something—whatever it is—it is like donating two cents to challenge a multi-trillion-dollar system, so one could easily get cynical and say, how absurd!

But then, there is that quiet smile that comes, almost unnoticed, when we say, "Yes, I know this, but I'm going to do something anyway." It is a smile that tells us that, while we are indeed within huge and powerful systems, we are not merely the product of them. Bella thinks that is what the prophetic tradition[13]

---

[13] Cornel West also describes the prophetic tradition in *Prophetic Fragments: Illuminations of the Crisis in American Religion & Culture* (1993).

is about, a history of acts seen as absurd, trivial, unrewarding, risky, and at times terribly dangerous, but sustained nevertheless by a spirit expressed through a quiet smile so subtle that it is only understood if you act and then experience it. And where will this smile lead us? He does not pretend to know. He says, "listen to the wind; you are not sure where it is coming from or going to, but you know its presence. Smile and do something."

Bella's Reframing of *Salmon 2100*

Given the above objections and obstacles, Bella claims a radical reframing is necessary. In *Salmon 2100*, the problem is framed in a linear way: preferences directly determine the future. But there is a something wrong here, according to Bella. The world is nonlinear, and this changes everything!

> Of course, if we limit our view, localize it in time and space, linear presumptions usually work quite well. In a similar manner, we can assume the world to be flat if we don't look very far. But, the world is neither linear nor flat. And, in both cases, what works on small scales (linear and flat earth thinking) becomes a dangerous form of blindness when applied to larger scales. Legacy is a matter of larger scales! Nonlinearity matters! The world is neither flat nor linear (D. A. Bella, personal communication, October 13, 2009).

In *Salmon 2100*, however, the problem is framed in linear

terms, according to Bella. He goes on to state that "linearity is embedded (taken for granted) in the language, particularly in the presumptions, that relate preferences to outcomes" (Bella, personal communication, October 13, 2009). On small scales – like shopping in a supermarket – such linearity seems reasonable and useful, but when it is applied to large scales (our legacy in 2100), linearity is a form of blindness, with serious and even catastrophic consequences.

Core Policy Driver #4

In response to Lackey et al.'s framing of the salmon "crisis," Bella agreed with the assessment that more of the same kinds of restoration attempts will not suffice. He instead proposed "A Wild Salmon National Park" (Bella, 2006, p. 132), but critics dismissed the proposal as unrealistic. This dismissal led Bella to reframe the problem, especially core policy driver #4. He states:

> We appear to face a disturbing choice: we can either
> be unrealistic and hopeful or realistic and cynical.
> In the prophetic tradition, this would be called
> bondage. And the prophetic tradition is about
> liberation from bondage, particularly bondage of
> long-established modes of thinking (Bella, 2006, p.
> 133).

In face of the aforementioned "bondage," Bella began to reformulate the problem. He began raising new questions, new possibilities, and to regard the problem from new angles. In so doing, he attempted to change what is 'realistic' and, in the language of this book, is thereby radically reframing the problem.

This radical reframing proposes a change in "how to think about what we are doing" (Bella, 2006, p. 126). It focuses on reframing Lackey et al.'s core policy driver #4 which states:

> Individual and collective preferences directly determine the future of wild salmon, and substantial and pervasive changes must take place in these preferences if the current long-term, downward trend in wild salmon abundance is to be reversed (Lackey et al., 2006, p. 66).

According to Lackey et al., this core driver "is perhaps the most obvious and arguably the most important" (Lackey et al., 2006, p. 66). Bella disagrees. Instead, Bella intends to show that the real driver involves a common misperception embedded in our language, practices, and institutions. "In brief, the misperception is that we can make linear presumptions in a world that is nonlinear."

> Linear presumptions mislead us to assume that if the parts (individual actions) are acceptable at the time taken, then the whole that emerges (from all actions) will also be acceptable (Bella, 2006, p. 133).

Because the world is mostly nonlinear, we must extend our concerns beyond limited domains where the world appears to be linear (Bella, 2006, p. 136). Therefore, Bella radically reframes core policy driver #4 to read:

> Linear presumptions—embedded in our
> institutions, language, and practices—distort our
> perceptions, misdirect our actions, and undermine
> our sense of responsibility, setting the stage for
> emerging outcomes that are contrary to our highest
> ideals, our most treasured values, and the faith
> tradition that many of us hold dear. The decline of
> wild salmon is a symbol of such emergent
> outcomes. Radical changes in these presumptions
> are necessary (Bella, 2006, p. 134).

In summary, when Bella proposed a wild salmon national park, he met dismissal. His proposal was viewed as 'unrealistic.' So, in response, he began reframing core policy driver #4 and encouraging us to question our legacy. In this way he is radically reframing the problem thereby changing what is "realistic."

Salmon as Prophetic Symbols

When assigned the question, "what is it *really* going to take to have wild salmon populations in significant, sustainable numbers through 2100?", Bella reframed the salmon as a prophetic symbol that points beyond itself to ask us about ourselves, the way we are living, and the legacy we are leaving. This difference in the framing of salmon as measurable objects of scientific study versus the reframing of salmon as prophetic symbols has great consequences on the strategies we employ. "The beleaguered salmon serve as a prophetic symbol calling us to expand our imagination, reframe the questions we have asked, and take actions to break out of established paradigms, models, ideologies, and institutionalized habits" (Bella, 2006, p. 128).

In his discussion, Bella contrasts the priestly and prophetic traditions. The former, which sustains authoritative bodies of knowledge best known by experts and administered through institutions, and the latter, which involves people who are neither experts nor supported by institutions; people who say "We need to open our eyes, expand our imagination, and change our ways" (Bella, 2006, p. 128).

Legacy

In review, Bella reframed the salmon as prophetic symbols of the wild. The key difference between Lackey's frame and Bella's reframe is in the way they define salmon. Lackey has framed the salmon as fish-to-be-counted. Bella has framed the salmon as a symbol that points beyond itself to ask us about ourselves, the way we are living, and the legacy we are leaving. This difference in the framing of salmon as measurable fish (objects of scientific study) versus salmon as symbols has major consequences for the strategies we employ.

One reviewer reminded Bella that the single question of the *Salmon 2100* was "what is it going to take to have wild salmon populations in significant, sustainable numbers through 2100?"

The reviewer then noted that his chapter offers little concerning measurable outputs, and numbers that could be used as standards of success. Why? (Bella, 2006, p. 139).

Bella replies to this critical question by first stating a common view of management strategies:

> In order to have a successful management strategy, one must establish measurable performance standards (e.g., size of fish populations) that serve to evaluate the success or failure of action (Bella, 2006, p. 139).

In contrast, Bella states that his chapter is based on a counterclaim.

> When the strategic standards for success are defined in measurable outcomes, the stage is set for the technological fix; technological solutions such as hatcheries are all but assured (Bella, 2006, p. 139).

In the next section, I discuss the ramifications of Bella's counterclaim. Bella's reframing leads to a different treatment of the problem.

Different Frames Can Lead to Different Treatments

One of the key differences that the two different frames in the case study make is in the kinds of treatments that they entail. Following Ackoff's (1999; 2006) typology of the treatments of problems, Lackey's frame implies a "solution" to the wild salmon problem, whereas Bella's reframing would imply a "dissolution." In other words, when Lackey et al. discuss the nature of the problem as basically being a matter of economic trade-offs – especially in core policy driver #4, the resultant treatment presumes that the context in which these trade-offs occur is a given. With the focus on individual trade-offs being summed into collective values and preferences, the reader is led to not even consider the context. Instead, the focus is on individual trade-off decisions and individual policy prescriptions. Due to this framing of the problem setting, we are led to seek a limited "solution" that assumes the context as given as opposed to a "dissolution" that seeks to redesign the context itself.

In contrast, Bella's reframing of the problem, encourages a multipurpose dissolution that questions how the contexts of economic trade-offs themselves may emerge unpredictably, the recent financial meltdown being a case in point. Thus, we may affect not merely trade-off decisions, but also the contexts in which trade-offs occur. Due to the emergent nature of nonlinear systems, we may not be able to predict exactly how contexts will change, but we can at least acknowledge that their existence can influence the probabilities that some contexts will emerge rather than others.

Thus, rather than limiting possible "solutions," Bella is encouraging us to seek *hopeful* dissolutions. This is not the same as a naïve optimism; we can still look at the "facts," but the way we frame and interpret the "facts" may change what we deem is possible. Instead of succumbing to "doom and gloom" when looking at the wild salmon problem, Bella asks us about the legacy we want to leave, not just numbers of salmon, but their habitat—the wild salmon national park that will allow humans to seek a multi-purpose dissolution to the problem.

In sum, Bella warns that the reductionistic linear presumptions that dominate the current socio-political context and the consequent framing of problems, limit our perceptions of what is realistic. Given this context, Bella asks, "What actions can we take so that the legacy we leave to our children and their children will be honored rather than lamented? (Bella, 2006, p. 10).

In summary, there is clear evidence that Bella has radically reframed Lackey et al's expansive reframing of the wild Pacific salmon problem. While Bella agrees with three out of the four core policy drivers, he specifically disagrees with policy driver number four, the very driver Lackey claims is the most obvious lesson learned from the project. As Lackey writes under Lesson #4— Individuals Select from Among Desirable Alternatives, "A straightforward lesson learned was that for wild salmon and other diadromous species, our individual and collective preferences directly determine their future, and substantial and pervasive changes must take place in these preferences if current long-term downward trends are to be reversed" (Lackey, 2009, p. 614). Lackey underscores this statement by declaring that "This lesson learned is perhaps the most obvious and arguably the most important" (Lackey, 2009, p. 614).

However, in his radical reframing of the problem, Bella rejects this specific driver:

> I intend to show that the real driver involves a common misperception embedded within our language, practices, and institutions. In brief, the misperception is that we can make linear presumptions in a world that is nonlinear. Linear presumptions mislead us to assume that if the parts (individual actions) are acceptable, at the time taken, then the whole that emerges (from all actions) will also be acceptable (Bella, 2006, p. 133).

So, why did Lackey not include Bella's lesson in his paper on lessons learned from the project? It could be that this is a confirmation of Lakoff's observation that "If the facts do not fit the frame, the frame stays and the facts bounce off" (Lakoff, 2004, p. 17). This lesson not learned then, seems to be a confirmation of Lakoff's claim and a confirmation that the way a problem is framed affects what is realistic. Bella clearly imagines a kind of wild salmon national park that he views as realistic based on his framing of the problem, but in Lackey's framing of the problem that takes current institutional and political realities as "givens" and insists on producing sustainable numbers of wild salmon by 2100 as the "required," Bella's wild salmon national park proposal and his radical reframing could easily be viewed as unrealistic and therefore dismissed as a lesson not learned.

CHAPTER 6

A Discussion on Reframing

In this discussion, as in the entire book, the term 'frame' is intended to invoke the metaphor of the picture frame, whereas the word 'framework' refers to the metaphor of the scaffold. The academic literature has been largely directed to frameworks and the radical reframing of frameworks (worldviews, paradigms, mental models, etc.) rather than also directed to institutionalized frameworks – which act as 'frames.' Thus, despite such literature, the *structure* of universities and other institutions continue to reinforce specific frames—borders that focus attention upon that which is enclosed. Arguably, the ability to teach and do research across these borders has become more difficult unless outside agencies provide funding. Even then, funding agencies impose and enforce frames as defined in funded proposals. Thus we find that while academic papers and books do address different frameworks, often they do so within the context of institutionalized frameworks.

As an example, the management literature includes a number of radically different frameworks (for example, Ackoff, Magidson, & Addison, 2006) However, this literature has done little to open up opportunities to transcend the institutionalized frameworks — frames that take as given the borders of schools, departments and disciplines—of the university itself. Nor has this literature done much to describe, much less challenge, the frames imposed by funding agencies.

The academic discussion of frameworks has been largely confined by the borders of institutionalized frames. For example, the discussions of frameworks within business rarely cross over to environmental scientists. Likewise, discussions of frameworks to better understand ecosystems seldom cross over from environmental science to business. Therefore, environmental scientists and business professors rarely see much relevance in each other's discussion of frameworks

This study takes a different approach to reframing. It pertains to frames. It acknowledges the importance of frames in education, particularly the way that the framing of problems instills disciplined approaches to problem solving, which is a paramount purpose in the fundamental courses in science and engineering science. Then, it examines how this practice of framing, when institutionalized, can blind problem solvers to emerging problems with potentials for catastrophic outcomes. After *Salmon 2100* was completed, as stated earlier, Lackey wrote "Challenges to Sustaining Diadromous Fishes through 2100: Lessons Learned from Western North America" (Lackey, 2009). In this paper

Lackey discussed the lessons learned from the *Salmon 2100* Project. Bella read this paper and found no evidence that his radical reframing was a significant lesson learned for Lackey.

Bella's Reading of Lackey's "Lessons Learned"

Bella re-read (again) Lackey's "Lessons Learned" paper and he found nothing in it—positive or negative—that even hints at the alternative framing that he developed in "Legacy" (his chapter in *Salmon 2100*). In fact, Lackey states lessons learned from *Salmon 2100* that are opposite of what Bella emphatically stated in "Legacy."

To see what is at stake at a deep level—far beyond concerns for salmon—try the following exercise. First, take key sentences from Lackey's paper. Read them to make sure that they are central to his broader claims. That is, make sure that these sentences are not taken out of context. Then strip out the verbiage specifically directed to salmon and fisheries biologists. Write down and read the stripped down statements. They express fundamental claims.

As an example Lackey states,

> In discussions about the future of salmon, it is easy
> to find comfort in debating nuances of hatchery
> genetics, evolutionarily significant units, dam
> breaching, smolt barging, selective fishing
> regulations, predatory bird control, habitat
> restoration, and atmospheric and oceanic climate,
> and unintentionally mislead the public about the
> realities of the situation. As discomforting as it may

be to disclose the future of wild salmon and other diadromous species relative to society's apparent values and preferences, our most useful contribution as fisheries scientists is providing information and assessments that are policy-relevant but policy-neutral, understandable to the public and decision makers, and scrupulously realistic about the future (Lackey, 2009, p. 616).

Now strip-out the words specifically referring to salmon, hatcheries, etc. and fisheries biologists. One obtains the *realistic challenge*:

...it is easy to... unintentionally mislead the public about the realities of the situation... our most useful contribution... is providing information and assessments that are... scrupulously realistic about the future (Lackey, 2009, p. 616).

This stripped down quote states a problem and challenge that is more fundamental than anything we might say about salmon, hatcheries, national parks, etc. This realistic challenge is a paramount concern of "Legacy;" the wild salmon national park proposed in this chapter is only an example to illustrate these more fundamental concerns. Moreover, the frame presumption implicit in core policy driver #4 allows scientists to justify their work and, ironically, to "unintentionally" mislead the public. The following section clarifies this irony.

The Irony of Core Policy Driver #4

There is an irony revealed by Bella's radical reframing of Lackey et al.'s core policy driver #4. The irony rests on the reasoning of a conditional statement ("if...then..."). The conditional statement is as follows:

If:

1) Scientists presume as "fact" that outcomes arise from preferences with little or no mention of emergent systems in human affairs,

2) But emergent systems do in fact shape (mold, provide the context for) human behaviors including the behaviors of scientists

3) And the outcomes arising from such emergent systems can and do produce harmful and potentially catastrophic outcomes that few, if any, would prefer

Then scientists are:

4) "unintentionally misleading the public about the realities of the situation" (Lackey's words).

Moreover, scientists are "unintentionally":

5) allowing emergent systems to act covertly (overlooked, not mentioned, much less exposed in the "facts" that describe the "realities of the situation)"

6) And providing legitimacy to the outcomes of emergent systems by stating or implying that "preferences determine outcomes."

The irony that follows from this logic is that while these outcomes (outlined above) do not reflect, express, or arise from the preferences of environmental scientists, they *do* arise from the presumption that preferences determine outcomes. If, among environmental scientists, there is a presumption that this "lesson learned" is "perhaps the most obvious and arguably the most important" (Lackey's words), then a revealing irony is apparent: The behaviors of environmental scientists themselves violate the presumed "fact" that "preferences determine outcomes." Expressed another way, the presumption that enables a focused approach to problems enables the emergence of problems that do not fit the presumption. In short, the irony is that the scientists, in this conditional statement, are enabling outcomes that they do not prefer by presuming that outcomes come from preferences.

Radical Reframing As a Means to Transcending the Hopeless Dilemma

The results of the quantitative and the qualitative data provide evidence that Lackey et al. performed an expansive reframing, whereas Bella provided a radical reframing of the wild Pacific salmon problem. But does Bella's reframing really make a difference? Is what is considered realistic truly changed? To answer this question, it is useful to return to the apparent choice in the 'Hopeless Dilemma' (Figure 1) that has arisen, not only out of this the case study but in the environmental sciences, in general.

Does radical reframing allow a new frame to *transcend* the

"hopeless dilemma?" How can radical reframing do this? First, I posit the question: Is it possible to be both a 'realist' and an 'optimist?' Sometimes it seems the answer is, 'no.' If one is 'realistic' about the problems of climate change, population, severe poverty, loss of biodiversity, energy, food, public health, economic globalization, and toxins in the environment, then the evidence does not warrant much room for optimism. So, are these the only options: Delusional optimism or doom and gloom (pessimism)? This is a false dichotomy.

Firstly, the kind of thinking illustrated in Figure 1 implies that we have a one dimensional choice. See Figure 17.

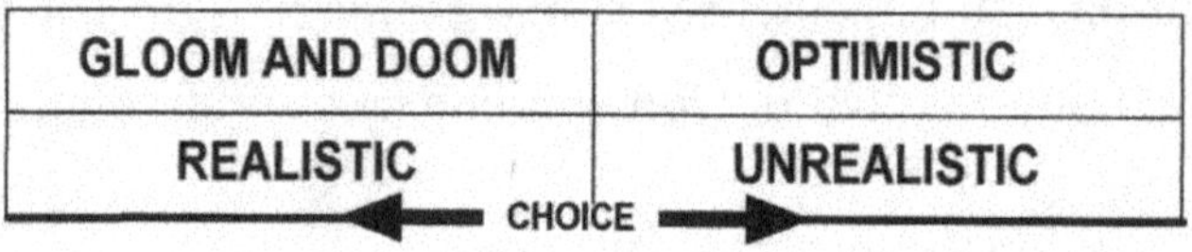

GLOOM AND DOOM	OPTIMISTIC
REALISTIC	UNREALISTIC

← CHOICE →

Figure 17. The Implied One Dimension of the Choice in the Hopeless Dilemma

Upon further consideration, however, this implied that the one dimension of choice in Figure 17 may be two dimensional if we add a new dimension (the vertical axis) for what is realistic and unrealistic, as in Figure 18.

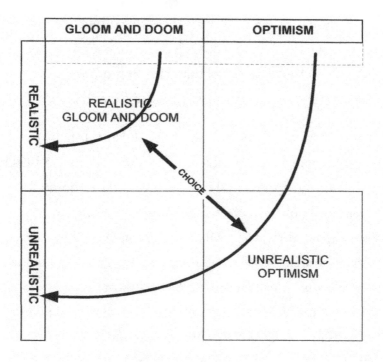

Figure 18. From One Dimension to Two Dimensions

By adding this vertical dimension, one can see that it may be at least theoretically possible to have a new choice between realistic optimism and unrealistic gloom and doom (pessimism). Look at the open spaces that are opened up. It is hard to even talk about these spaces unless the problem is radically reframed. The radically new choice is orthogonal to the first choice between realistic gloom and doom and unrealistic optimism. It is hard to comprehend this radically new choice if you are measuring based on the orthogonal choice just as you cannot measure the height of the ceiling by taking measurements of the floor. Translating between these two different choices is a challenge. As a frame translator, I have translated between these two frames via the frame analysis methods in this book. Now, refer back to Figure 18. It is itself a visual translation tool that will allow one to see the new choice and the new open spaces of possibilities that exist.

The way in which a problem is framed affects this new dimension and many authors have tried to offer realistic optimism by reframing the environmental problems with the notions of progress and technological optimism. These kinds of views challenge us to not dwell so much on the problems of the present, but to look at how far we have come through progress, and to look at how more people means more minds for innovation and technology that could solve our problems. Environmental authors Ted Nordhaus and Michael Shellenberger are the chairman and president of the Breakthrough Institute (retrieved from Internet 1/6/10 at http://www.thebreakthrough.org/) which is a special project of the Rockefeller Philanthropy Advisors, Inc. They wrote *Break Through: From the Death of Environmentalism to the Politics of Possibility* (2007) in order to reframe environmentalism. They argue that environmentalism must move from a politics of limits to a politics of possibilities.

> Few things have hampered environmentalism more
> than its longstanding position that limits to growth
> are the remedy for ecological crises. We argue for
> an explicitly pro-growth agenda that defines the
> kind of prosperity we believe is necessary to
> improve the quality of human life and to overcome
> ecological crises (Nordhaus & Shellenberger, 2007,

p. 15).

Break Through (2007) represents an attempt to reframe environmentalism in a realistic and optimistic way. The way is dependent on a pro-growth aspirational model that focuses on investment in technological research and development in order to solve our current ecological crises.

Former University of Maryland Business Professor Julian Simon (cited by Bella on p. 136 in "Legacy") offered an additional example of an attempt to reframe our current predicament in a 'realistically' optimistic way. Simon was the primary proponent of the cornucopian belief in endless benefits from resources and unlimited population growth empowered by technological progress, innovation, and recycling. His 1981 book, *The Ultimate Resource*, is a criticism of the conventional wisdom on population growth, raw-material scarcity, and resource consumption in the modern world. Simon argues that notions of increasing resource-scarcity ignore the long-term declines in wage-adjusted raw material prices. Viewed economically, he argues, increasing wealth and technology make more resources available; although supplies may be limited physically, they may be viewed as economically indefinite as old resources are recycled and new alternatives are developed by the market. Simon challenges the idea of a pending Malthusian catastrophe—that an increase in population has negative economic consequences; that population is a drain on natural resources; and that we stand at risk of running out of resources through over-consumption. Simon argues instead that population is the solution to resource scarcities and environmental problems, because people and markets innovate.

What if arguments like *Break Through* and *The Ultimate Resource* do not convince a person that we will sustain our life support systems in any kind of honorable way? What if one sees that growth is part of the problem, not the solution? What if one does not see convincing proof that more humans equals more intelligence and more long-term life support saving qualities? If one does again see limits to growth and technology as a necessary part of the solution, then one might end up back in the view that the only realistic fate is gloom and doom. We may again succumb to depression and despair as the only viable sentiments when we realistically look at the complex environmental problems. Perhaps we cannot be truly optimistic based on the evidence. This is not the best of all possible worlds. Maybe we need an alternative economic model.

There is an alternative to Simon's cornucopian model that is offered by the Center for the Advancement of Steady-State Economics (CASSE). The center was created to spread the word of American ecological economist Herman Daly's conception of a "steady-state" economy. Herman Daly (who was Senior Economist in the Environment Department of the World Bank) said that "CASSE is the foremost organization in advancing the precepts of the steady state economy to citizens and policy makers – an indispensable resource" (Taken from internet on 12/8/09: http://www.steadystate.org/). According to CASSE's website "The key features of a steady state economy are: (1) sustainable scale, in which economic activities fit within the capacity provided by ecosystems; (2) fair distribution of wealth; and (3) efficient allocation of resources" (Taken from internet on 12/8/09: http://www.steadystate.org/CASSEBasics.html). CASSE President Brian Czech and Herman Daly wrote In "My Opinion: The Steady State Economy—What it is, Entails, and Connotes" (Czech & Daly 2004) in order to describe and define what is truly meant by a "steady state economy" and how it is different from the pro-growth models as advocated by Simon and others. In summary, CASSE offers a counter argument to Simon's pro-growth technological optimism arguments. So why don't we hear more about Herman Daly's views? Why have such alternative views so often been dismissed as unrealistic and given the 'kiss of death?' I contend in this book, that the way we frame our environmental problems affects our view of what is realistic and unrealistic. So, maybe a radical reframing of our economy must occur before we can begin to see views such as Herman Daly's steady-state economy as realistic. If we do not reframe the economy and instead continue with 'business as usual,' then it becomes very difficult to be 'realistically optimistic.'

If this is the case, we may need to abandon the notion of optimism altogether and instead pursue hope. After all, the categories of optimism and pessimism are limited and limiting. It may not be possible to have optimism when one considers the evidence of the 'wicked' environmental problems, but one can still have realistic hope. Hope is not the same thing as optimism. Optimism is the belief that, based on the evidence, trends, etc., there is a good probability that the problems will be solved. Hope is different. It is the courage to do the right thing because it makes sense regardless of the outcome. Cornel West, graduate of Harvard and Princeton, has spoken eloquently about the distinction between hope and optimism. When asked by *Rolling Stone* if he was optimistic about the future, West states,

> The categories of optimism and pessimism don't exist for me. I'm a blues man. A blues man is a prisoner of hope, and hope is a qualitatively different category than optimism. Optimism is a secular construct, a calculation of probability (West, 2007).

West then defines hope,

> Hope wrestles with despair, but it doesn't generate optimism. It just generates this energy to be courageous, to bear witness, to see what the end is going to be. No guarantee, unfinished, open-ended, I am a prisoner of hope. I'm going to die full of hope. There's no doubt about that, because that is a choice I make. But at the same time, the end doesn't look too good right now (West, 2007).

Elsewhere, West (1993) again spoke eloquently on this distinction:

> Last, but not least, there is a need for audacious hope. And it's not optimism. I'm in no way an optimist. I've been black in America for 39 years. No ground for optimism here, given the progress and regress and three steps forward and four steps backward. Optimism is a notion that there's sufficient evidence that would allow us to infer that if we keep doing what we're doing, things will get

better. I don't believe that.

West then goes on to define "audacious hope:"

> I'm a prisoner of hope, that's something else.
> Cutting against the grain, against the evidence.
> William James said it so well in that grand and
> masterful essay of his of 1879 called "The Sentiment
> of Rationality," where he talked about faith being
> the courage to act when doubt is warranted. And
> that's what I'm talking about.

With this distinction in mind, can we be realistically hopeful? Even if we cannot be optimistic with "sufficient evidence that would allow us to infer that if we keep doing what we're doing, things will get better," can we have hope that "cuts against the grain, against the evidence?" Can we have the "courage to act when doubt is warranted?"

Firstly, consider if the answer to these questions is "no." Then we are back to doom and gloom, which often leads to depression, and depression often kills the motivation to do anything when, after all, "it won't make a bit of difference." Secondly, let us consider if the answer to these two questions is "yes." Then we can be "realistic," we can "face the facts" and still find the courage to act.

Consider Figure 19 for a moment.

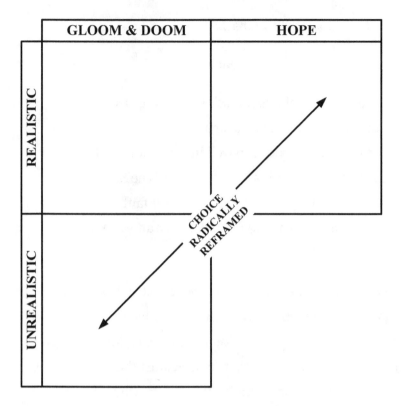

Figure 19. Choice Radically Reframed: Hope is Not Equal to Optimism

So, how can one have 'realistic hope?' I suggest that we have to ask ourselves how we know what is realistic and unrealistic. We must look at how the environmental problems have been framed, for the way one frames problems affects what one considers realistic and unrealistic problem treatments. As Hubert Dreyfus, professor of philosophy at the University of California at Berkeley, has pointed out:

> The representation that defines the problem space
> is the problem solver's "way of looking at" the

problem and also specifies the form of solutions. Choosing a representation that is right for a problem can improve spectacularly the efficiency of the solution-finding process. The choice of problem representation is...a creative act (Dreyfus, 1980, p. 18).

So when we face crises and dilemmas where we see no realistic way of being optimistic, we must first ask how we have framed the problem. How is the 'problem space' represented? Next, we must radically reframe the problem with new representations of the problem space if we are to find realistic hope.

Recall that in Lackey's lessons learned paper he states that,

> ...several reviewers suggested that if my objective in writing was to help save wild salmon (it was not), then the accurate, realistic message would leave proponents dejected. This common sentiment is captured by the following: You have to give those of us trying to restore wild salmon some hope of success (Lackey, 2009, p. 616).

Perhaps the way to realistic hope is in radically reframing Lackey's framing of the problem. Bella did this, and he stated in his "Legacy" chapter in *Salmon 2100*,

> But if our society endures, is it realistic to hope that a wild salmon national park [*sic*] could emerge by 2100? And if we cannot realistically hope for this outcome, how can we realistically hope for other

outcomes that would constitute an honorable legacy? There is much more involved in such questions than the survival of fish runs (Bella, 2006, p. 140-141)!

His answer is that it is realistic to hope for such things, but hope can only be had if the problem is radically reframed. (See the conclusion for the effects of the radically reframed choice illustrated in Figure 19.)

Discussion of Reframing in *Salmon 2100*: The Case Study

Salmon 2100 (Lackey et al., 2006) provides evidence that we need to examine the ways in which we frame problem settings. As with other 'wicked' problems, there are many ways of looking at the problem, many paths worth exploring, and rarely is there only one "right" solution (Bardwell, 1991). The risks are high, and the consequences of our actions are potentially long term and irreversible. In addition, as pointed out earlier with expansive and radical reframing, many scientists may take a "whole systems" view of ecosystems, but then be reductionistic when human institutions are involved.[14]

After exploring the different frames in *Salmon 2100* through frame analyses and interviews with Bella, Lackey, Lach,

---

[14] This is a key difference between scientific research and science policy as observed in *The Honest Broker* (Pielke, 2007).

and Duncan, I found the "Frames and Reframes Diagram" in Figure 20 to be consistently useful. In addition, I found a need for a method of revealing self-reinforcing emergent contexts that Bella describes. First, I will explain "The Frames and Reframes Diagram" in Figure 20.

Frames and Reframes Diagram

Who/What We Think

		Them	Us
How We Think	**Nonlinear**	2	4
	Linear	1	3

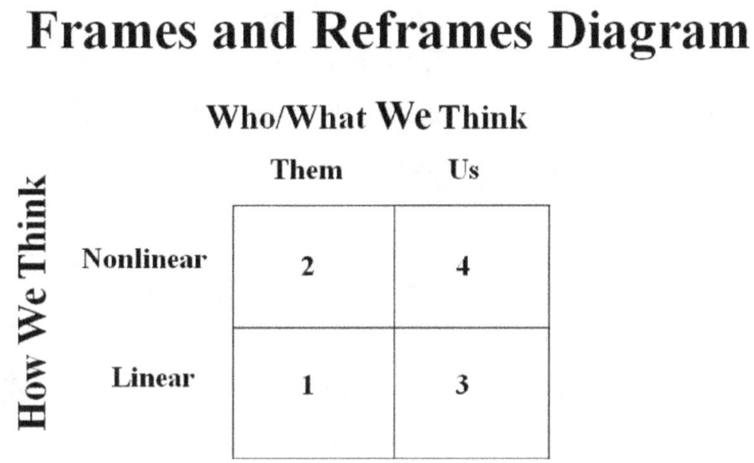

Figure 20. How do we approach problems in a realistic way?

Each box in the above frames and reframes diagram (Figure 20) represents a different perspective of the problem. In region 1 we tend to frame problems using linear logic, and we focus on 'them.' If we reframe the problem to region 2, then we acknowledge that the ecosystem of the fishery is complex (nonlinear) and there are many interactions. While this frame may be more comprehensive, it is often not viewed as very practical precisely because the ecosystem is so complex.

If we frame the wild salmon problem according to region 3,

then we acknowledge that we are including ourselves in the frame. We are not just talking about 'them.' In order to understand our connection to the problem, we use linear methods of research. Often we explore our values and preferences through surveys and questionnaires. We then use statistics to find averages, probabilities, and tendencies. Generally, however, we are adding up the values and preferences of individuals in order to understand the 'whole system'. Region 3 tends to lead to the conclusion that, until people change their values and preferences, nothing will change because the system is merely the sum of the parts.

Finally, if we reframe the problem yet again according to region 4, then we frame the problem in a nonlinear way that acknowledges emergent phenomena in both 'natural' and 'social' systems. We accept that the whole is not merely the sum of the parts, i.e., emergence. In addition, we acknowledge that "we are talking about us," so there is some sense of self-reflection and responsibility. We are not only looking to blame 'them' or count 'them,' i.e., society, policymakers, or the public. The individuals in the MAHB do this to some extent when they blamed society. In 2009, the MAHB wrote, "Yet society stubbornly refuses to take comprehensive steps to deal with them (the environmental problems) and their drivers,

- population growth
- overconsumption by the rich, and
- the deployment of environmentally malign technologies" (Mission, n.d., para. 2).

In region 4 we are looking instead to take some responsibility for the system we are in. We can look for ways to represent that system (context sketches) and then transcend it. When we can see what we are talking about through visual representations, we can move towards new beginnings, new designs, new possibilities, and new hopes.

In summary, we can move from regions 1 to 2 through reframing to include complexity and nonlinear reality, but we are primarily focused on other people, other fish, other systems, or simply 'them'. The responsibility and blame is with 'them'. Similarly, we can move from region 3 to Frame 4 through reframing. However, in order to see the nonlinear emergent systems, we need a representation. Bella's sketches provide such representations. The sketches are simplifications, as are all representations, but they give us a sense of the patterns that are persistent over time in complex, adapting, nonlinear (CANL) systems, which tend to share the following characteristics:

- They have many components.

- Each component is directly linked to (influenced by) only a few other components (a tiny fraction of the total).

- The links between components form vast networks through which multiple pathways of influence and exchange can be traced.

- These networks contain multiple loops of influence and exchange which can dampen or amplify deviations (negative or positive reinforcements) in nonlinear ways.

- The formation of these networks is an adaptive process that involves the interplay of order and disorder over time (i.e., structure emerges from a history of tests, challenges, and influences).

- The adaptive tendency is toward mutually reinforcing networks of influence and exchange that serve to prevent the growth and spread of disorders and coordinate the activities of many components far beyond the direct influence of any component.

- This systemic coordination leads to "emergent" behaviors, outcomes, and capabilities that cannot be reduced to the behaviors of components (Bella, 1992).

Us Versus Them and "Humane Sciences"

In addition, to reframing from linear to nonlinear or vice versa, one can reframe problems from talking of an 'it' to talking of a 'they.' This, in turn, can become a 'we' talking of a 'they'; then a 'we' talking of 'you.' This can then become 'we' talking 'with' you; and finally—the goal—'we all' talking together about 'us' (Smith, 1997, p. 144).[15] Smith was professor emeritus of the Comparative

---

[15] There are passages in Freire's *Pedagogy of the Oppressed* (1970) that are very similar. This current of thought is present in contemporary pedagogical models.

History of Religion at Harvard University. His characterization of the study of comparative religion can be useful in framing and reframing scientific discussions and explorations.

Indeed, Smith refers to the application of such a reframing as the "Humane Sciences" in his Chapter 8 "Objectivity and the Humane Sciences: A New Proposal" (Smith, 1997, p. 121). He describes the humane sciences as follows:

> We are talking about the study of persons by persons. By corporate critical self-consciousness I mean that critical, rational, inductive self consciousness by which a community of persons, constituted at a minimum by two persons, the one being studied and the one studying, but ideally by the whole human race, is aware of any given particular human condition or action as a condition or action of itself as a community, yet of one part but not of the whole of itself; and is aware of it as it is experienced and understood simultaneously both subjectively (personally, existentially) and objectively (externally, critically, analytically; as one used to say, scientifically) (Smith, 1997, p. 124).

Following this corporate critical self-consciousness, one person may study another 'objectively,' but then they verify their observations not just by comparing and contrasting with another 'objective' observer, but they also add the experience of the subject during the observation as valid information. Smith writes "No statement involving persons is valid, I propose, unless its validity can be verified both by the persons involved and by critical observers not involved" (Smith, 1997, p. 125). This, of course, means that, the researcher must listen carefully and empathically in order to understand the view of the subject: "The proper goal of humane knowing, then, the ideal to which we should aspire academically, scientifically, is not objectivity but corporate critical self-consciousness" (Smith, 1997, p. 125).

Simplicity, Organized Complexity, and Radical Reframing

Recall Warren Weaver and the need for further study into systems of 'organized complexity.' While reading Weaver's article, as discussed in the section on Radical Reframing, it prompted me to revise my "Frames and Reframes Diagram."

If we revisit the "Frames and Reframes Diagram" now after considering Weaver's argument we may change the diagram slightly to refer to 'emergent wholes' vs. 'parts' on the vertical axis. The 'emergent wholes' would include Weaver's call for studies of 'organized complexity' and what we called the 'nonlinear' in the previous diagram. In addition, the 'parts' would correlate with studies of 'simplicity' and the 'linear' way of thinking. This modified version of the Frames and Reframes Diagram may be the most readily understandable language that illustrates the "radical reframing" that is discussed in this book.

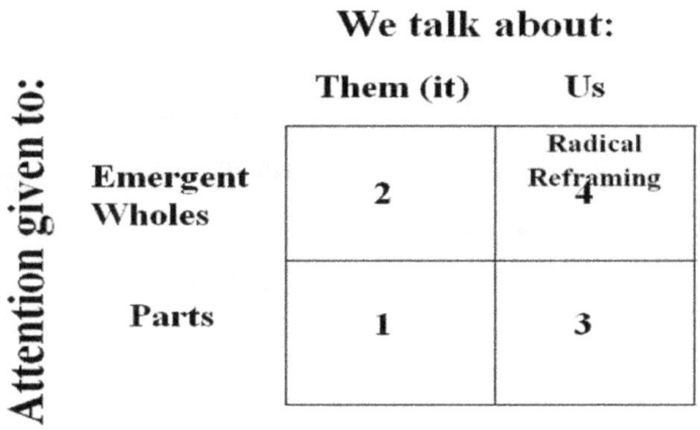

We talk about:

	Them (it)	Us
Emergent Wholes	2	Radical Reframing 4
Parts	1	3

Attention given to:

Figure 21. Revised Frames and Reframes Diagram

General Discussion of Radical Reframing After the Case Study

Now, explore the following map/diagram and the urban sprawl context (saying 'therefore' when reading forward on an arrow and 'because' when reading backward).

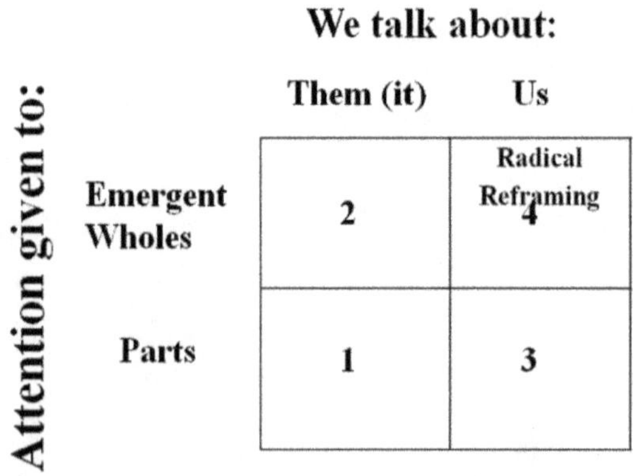

We talk about:

Attention given to:	Them (it)	Us
Emergent Wholes	2	Radical Reframing 4
Parts	1	3

Figure 22. Map of The Frames and Reframes Diagram (Radical Reframing occurs in region #4)

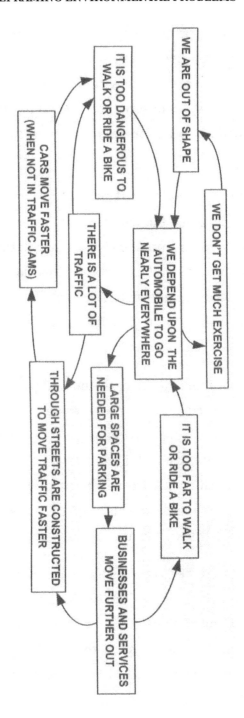

Figure 23. Sketch of Urban Sprawl Context

Note that when people read one of these maps—sketches, diagrams—they draw upon their own experiences, not the authority of 'them,' the experts. A problem is reframed when they come to see that they themselves are caught up within a context and their 'in context' behaviors lead to consequences that none of them would prefer.

If the context is allowed to define what is 'realistic,' then there is no 'realistic' hope that these unpreferred consequences can be avoided. However, if the problem is reframed so that the context itself is the problem, then the question arises, do contexts change in linear (predictable, cause and effect) ways? If 'hope' is based upon knowing with certainty that 'if I do A, outcome B will occur,' then there is no 'hope.'

Radical reframing must face the unpredictability of contextual change. But one need be neither helpless nor hopeless, that is, we are not prisoners of context. We can act in ways that might help bring about favorable contextual shifts that lead toward a more honorable legacy. But such acts will be out of context from perspectives within the established context. Without this radical reframing, such ways will be dismissed as "unrealistic."

The sketches serve to expose the character of contexts that both shape our behaviors and lead to outcomes that nobody would prefer. The concept of 'Kyros time' was introduced by Bella to radically reframe how we can act both realistically and hopefully. This radical reframing is described as prophetic rather than priestly. That is, the reframing does not cite the authority of experts; we do not talk about 'them,' the experts. Instead, the reframing says, "open up your eyes; look at what you are caught up within and look at the consequences." In other words, 'we do talk about us.'

Because the notion of Kyros time is so critical to understanding Bella's radical reframing of the problem, I will elucidate it here. White (1987) elegantly summarizes the more extensive work of Onians (1973) to illustrate the origins and meanings of *kairos* (spelled Kyros in Bella's chapter), *"Kairos* is an ancient Greek word that means 'the right moment' or 'the opportune'" (White, 1987, p. 13). This word is often contrasted with *khronos (chronos)*, which is the measured time that we use every day on clocks and calendars, and which is the origin of our words such as chronology, chronometer, and synchronous. While it is true that every *kairos* time has a corresponding *khronos* time, that is not what is important to understanding *kairos*. The right or opportune moment happens when it happens, in its own time, and not in entirely predictable ways. In order to explain this conception of time, the ancient Greeks used at least two different stories. "In archery, it refers to an opening or 'opportunity' or, more precisely, a long tunnel through which the archer's arrow has to pass" (White, 1987, p. 13). Successful passage of a *kairos* requires that the archer's arrow be released not just accurately, but with enough power for it to penetrate the tunnel all the way to the target (which could be moving!). The other meaning of *kairos* ('the right moment') traces its origins to the art of weaving. In weaving, *kairos* is the 'critical time' when the weaver must draw the yarn through a gap that momentarily opens in the warp of the cloth being woven. "Putting the two meanings together, one might understand *kairos* to refer to a passing instant when an opening appears which must be driven through if success is to be achieved" (White, 1987; p. 13; Onians, 1973). With this conception of *kairos* in mind, we can better understand and interpret Bella's radical reframing. The concept of *khronos* time alone is too limiting for understanding Bella's radical reframing that includes notions of nonlinearity and emergent whole systems. So, with the aforementioned meaning of *kairos* in mind, we can better understand how the contexts which are modeled in Bella's context sketches can shift or be changed in the right or opportune moment.

For example, imagine that a group of people did read through Figure 23, the urban sprawl pattern. Imagine that they could see how they themselves were caught up within it. Imagine that they consider alternatives and some actions, however small, that they might take to change the context. This is the reframing of region 4 on the map in Figure 22. 'We talk about us not them' and 'we talk about whole patterns, not parts.'

With this kind of radical reframing in mind, one can appreciate why the extensive wild salmon national park outlined in "Legacy" might not be 'unrealistic.' Within our current context of organizational divisions, this proposal probably is unrealistic. However, by reframing the problem to see this context itself as a problem—an established arrangement that prevents something good from happening—we can have realistic hope that the context can change in favorable ways in *kairos* time perhaps.

It is easy to dismiss such notions when they are not based

upon a reframing that is indeed radical. Bella reminds us to think of recent history. At the time *Salmon 2100* was proposed, George W. Bush was President, Alan Greenspan—the leading proponent of the free market—had gained the authority of a secular high priest. Housing was booming. Access to mortgages was expanding, making the 'American Dream' available to more people. Credit was widely available. The economy, led by the financial industry, was growing. The market system had worked so well that there were even calls to privatize social security and some public lands. Within that context, spending public funds for a public good such as a wild salmon national park would indeed be seen as absolutely unrealistic (D. A. Bella, personal communication, October 13, 2009).

But within two years after *Salmon 2100* had been published, a non linear shift (a crash!) had occurred. Vast sums, trillions of dollars, of public funds were being used to bail out banks and auto companies. Money was pouring in to stimulate the economy. Clearly the context had shifted. It would have been good if the notion of legacy had been taken more seriously so that more lasting good could have been passed on to our children's children.

Should contextual shifts be seen as opportunities for scientists? Yes, certainly. Scientists quickly took the opportunity to call for more science funding from the stimulus money. Moreover, scientists and mathematicians—known as 'quants'—played key roles in the development of financial models that led to the crash. And they are pointing to more opportunities in its aftermath.

According to Bella, there is something perverse when

scientists benefit from their own 'in context' behaviors and then cash in on the context shifts. They benefit from the recovery when the contexts crash but then dismiss as 'unrealistic' proposals that: 1) could leave a more honorable legacy but 2) are out of context from the perspective within the current context.

Salmon 2100 and *A Common Fate*

My results and discussion in this study have given me some insights regarding what has been called a "conspiracy of optimism" in fisheries and forestry. In *Salmon 2100* Lackey asks, "Is there some kind of 'conspiracy of optimism' that has overtaken the scientific process?" He then states, "if the technical experts are truly pessimistic, somehow that judgment is not being communicated and understood by decision makers and others responsible for implementing salmon policy" (Lackey et al., 2006, p. ix). Recall that Lackey's questions are based on his observations at a conference. "The atmosphere surrounding the conference, typical of nearly all salmon meetings, was a mixture of policy complexity and scientific uncertainty, overlaid with an informal, public veneer of optimism" (Lackey et al., 2006, p. ix). Lackey goes on to claim that "as always, the unspoken premise was 'if the experts could just solve the *scientific* challenges, or if we could just get sufficient money to do more of what we are *already* doing, salmon runs could and would be brought back to significant and sustainable levels' " (Lackey et al., 2006, p. ix). In contrast to this public conference during the day, the tone around the table in the hotel that evening was different. In the private context, the same people who had revealed a "public veneer of optimism" during the day expressed that the limitations to salmon recovery were not primarily scientific, instead they recognized that *"dramatic policy changes* must be implemented if the long-term downward trend in wild salmon abundance was to be reversed" in the Pacific Northwest and California. Lackey observes that many of the people involved with the conference were the same ones sitting around the table, but the "tenor of the two discussions were as different as night and day" (Lackey et al., 2006, p. ix). "It was almost as if two parallel worlds existed, one of a fairly positive, optimistic perspective about the future of wild salmon, the other a highly skeptical, pessimistic assessment of any recovery strategies under consideration" (Lackey et al., 2006, p. ix).

Table 2. Private Pessimism and Public Optimism

PRIVATE PESSIMISM	PUBLIC OPTIMISM
"highly skeptical, pessimistic assessment of any recovery strategies under consideration"	"fairly positive, optimistic perspective about the future of wild salmon"

Lackey follows his observations with the question, "Why this dichotomy... Is there some kind of 'conspiracy of optimism?'" (Lackey et al., 2006, p. ix). In an interview, Bella agreed with Lackey's observations that there may be some kind of "conspiracy of optimism."

Later in their introduction, Lackey and his colleagues, frame the premise of the *Salmon 2100* Project. "Thus, it is likely that society will continue to chase the illusion that wild salmon runs can be restored without massive changes in the number, lifestyle, and philosophy of the human occupants of the western United States and Canada" (Lackey et al., 2006, p. 3). With this as the premise, Lackey et al. banish both "delusional optimism and baseless pessimism" from the project and frame the problem by writing:

> We know and understand the direct causes of the
> decline of wild salmon numbers. The trajectory
> remains downward. Nothing will change unless we
> address the core policy drivers of this trend: the

rules of commerce, particularly market
globalization; the increasing demand for natural
resources, especially high-quality water; the
unmentionable human population growth in the
region; and individual and collective preferences
regarding life style. Do we, as a society, understand
the connections? Can we, and do we want to, turn
the ship around? (Lackey et al., 2006, p. 57).

So, while Lackey et al. banish delusional optimism and baseless
pessimism with their framing of the problem, they also mention a
'conspiracy of optimism.' What is a 'conspiracy of optimism?'

There are two books that are critical to understanding 'a
conspiracy of optimism' as it relates to the Pacific Northwest and
wild salmon recovery efforts: *A Conspiracy of Optimism* (Hirt,
1994) and *A Common Fate* (Cone, 1995). Environmental historian
Paul Hirt titled his book on forestry research, *A Conspiracy of
Optimism* (1994). Hirt describes how the filtering and distortion of
information that occurred in forestry agencies has given an image
of technological optimism in the future of forestry, when there
were really serious problems and limitations to technology.

In forestry, a faith in technology and progress as solutions for
our environmental and sustainability problems reigned in the
1940s to 1960s America (Hirt, 1994).

Foresters' psychological investment in the efficacy
of intensive management was so powerful that it
filtered every assumption and perception.

> Furthermore enough science backed the faith in
> technological mastery over nature that foresters
> could assert an empirical foundation—and therefore
> unquestioned legitimacy—to their beliefs (Hirt,
> 1994, p. 294).

This faith in the technological fix was genuine and it led to what Hirt calls "a conspiracy of optimism," which became the title of his book (1994). He goes on to say, "forest researchers for the most part were only asking the kinds of questions that would advance the conspiracy of optimism" (1994, p. 294). Furthermore, "Agency leaders consigned to marginality research that pointed to the flaws in the faith, or else they deferred judgment on problems until 'additional studies' could be made" (1994, p. 294). In addition to filtering research questions, actual researcher jobs were affected by the "conspiracy of optimism." Hirt states that when criticism of the timber program occurred in the 1970s (e.g., the clearcutting controversy), agency leaders stepped up their self-justification. Those who followed the program got promoted whereas those who did not follow the program were viewed as trouble makers and were quickly replaced or transferred. "If lower echelon land managers make life difficult for upper echelon agency heads, they risk losing their jobs" (Hirt, 1994, p. 295). In summary, many researchers struggled within their own agencies if they did not follow the faith in technology to make intensive management sustainable over time with innovations. If researchers within the agency questioned the timber "program," then they risked losing their jobs. This conspiracy of optimism affected the course of scientific inquiry by affecting the way problems were framed, thereby affecting what were considered 'realistic' solutions.

Another book that discusses a conspiracy of optimism is *A Common Fate* (Cone, 1994). In *A Common Fate*, science communicator Joseph Cone (with the Oregon Sea Grant since 1983) describes the experiences of salmon science researchers similarly. In 1973, Congress passed the Endangered Species Act. A new generation of wildlife biologists appeared on the research scene, reflecting broader concerns for the environment. Gordie Reeves was one of this new generation[16]. During his early research years, Reeves carried out a performance audit of salmon restoration projects from California through Alaska. After performing many of these audits, Reeves learned how the fisheries-enhancement game was played and he learned to be skeptical of claims of success: A lot of 'paper salmon' were created (Cone, 1994, p. 13). Nevertheless, Reeves didn't blame individual biologists, because "most efforts were well intentioned" (Cone, 1994, p. 13). One model of what was happening was provided by Dave Bella at the 1989 American Fisheries Society meeting. He described and modeled what he called the "systemic distortion" of information. This systemic distortion of information is often an emergent quality of organizations that filters out "troubling matters." "Organizations selectively produce and sustain information favorable to them, which frequently means looking out for those at the top of the pyramid" (Cone, 1995, p. 26).

---

[16] Read Appendix A of *Reframing Environmental Problems* for a summary of an interview with Reeves. The interview explains why he refused to participate in *Salmon 2100* because of the way that Lackey et al. framed the problem. In addition, Reeves shares a *kairos* time.

Favorable assessments have survival value and "contrary assessments tend to be systematically filtered out" (Cone, 1995, p. 26). The cumulative outcome of such distortions of information is "systemic distortion of information". The filtering is not always intended by individuals; instead, it becomes part of the normal organizational culture and context over time.

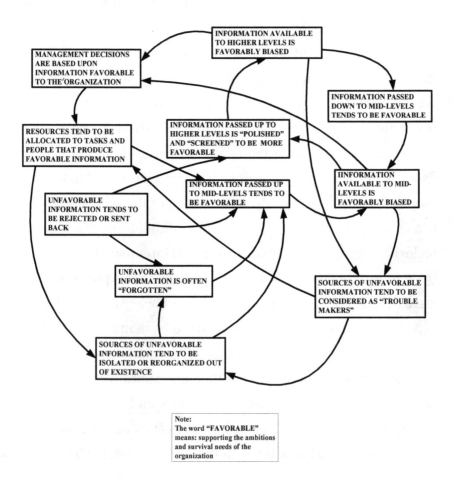

Figure 24. Bella's Systemic Distortion of Information Sketch Adapted from (Bella, 1996).

The marine biologist Jim Lichatowich then spoke at the same AFS meeting. Lichatowich had been assistant chief of fisheries for the Oregon Department of Fish and Wildlife for five years, but he had quit a year before due to policy differences with higher-ups in the department. "When a leader quits over a matter of principle, others who had been following him think hard about their value" (Cone. 1995, p. 27). Lichatowich tied the aforementioned forestry conspiracy of optimism with a similar phenomenon in fisheries. "Fishery managers once believed that both a fish agency and the fish resource could be protected by quiet inaction," he said (in Cone, 1995, p. 27). He went on to say that if the powerful Oregon timber lobby didn't like a new fishery regulation, the state agency would avoid the issue by not enforcing the regulation. The agencies were slow to challenge these special interests. He then argues that society changed, and new expectations were created for the agency. He stated, "As natural-resource managers and professionals, we have an important connection to future generations.... Our decisions will strongly affect the quality of life of our descendants—a big responsibility." He summed up by saying that, in carrying out that responsibility in a pluralistic democratic society, "...disagreeing with your neighbor is OK, disagreeing with the president's policies is OK, and even disagreeing with your boss is OK" (Lichatowich in Cone, 1995, p. 27).

Reeves made assessments similar to those of Bella and Lichatowich. He had come to believe that "agency officials worried about shifting political priorities that might affect agency budgets, and field staff busied themselves with technical activities that were driven by the prevailing political and economic priorities (Cone, 1995, p. 24). Little time or attention was left over for the larger and more troubling questions of values. "Avoidance of these questions had the effect, intended or not, of supporting the status quo" (Cone, 1995, p. 24). So, under a "conspiracy of optimism," the status quo was hard to change. In some cases, as with Lichatowich, people had to quit in order to speak of their experiences and values. Lichatowich eventually wrote *Salmon Without Rivers* (1999). He claims in that book that

> Euro-Americans saw a different landscape than the one the Indians had been living in with a high degree of harmony for several thousand years. They saw a fearful wilderness that had to be tamed, simplified, and controlled. They saw vast resources they needed to feed their voracious industrial economy. Their vision naturally directed them to reconstruct the Northwest to make it more like the places they came from than the place it was (Lichatowich, 1999, p. xiv).

The purpose of the book is to present the history of the impact of that vision on salmon. He argues that restoration efforts, though well funded, have failed. "They have failed because they are largely derived from the same worldview and assumptions that created the problem in the first place" (Lichatowich, 1999, p. xiv). Viewing this historical relationship between people and the salmon, especially over the last 150 years, "leaves little room for optimism" (Lichatowich, 1999, p. xiv).

In summary, there is a "conspiracy of optimism" that distorts and filters information that is favorable to an organization. In addition, the conspiracy of optimism leads to the hiring of people who are in agreement with the "optimism" of the organization, and the firing of people who do not conform with the policy preferences within the organization. The net effect is that when organizations tend to avoid the big ethical questions, they tend to maintain the status quo by filtering out disturbing information and disturbing employees. Sometimes, as in Lichatowich's experience, one has to quit their job in order to get out the message about controversial environmental problems. Individual researchers struggle to ask new innovative questions that are outside of the conspiracy of optimism. When employees question the limitations of technology and the possibility that we need to address larger issues, they often risk losing their job.

So, when Lackey et al (2006) ask if there is some kind of "conspiracy of optimism" that has overtaken the scientific process related to wild Pacific salmon recovery in *Salmon 2100*, there are some parallel historical developments described by Hirt. As described earlier, Hirt's *Conspiracy of Optimism* (1994) describes processes he observed in the context of intensive forest management in the Pacific Northwest. Granted, there is no true conspiracy in that there are no people intentionally trying to deceive. Rather, it is a *systemic* distortion of information which filters out pessimistic views/evidence of current strategies and instead favors views/evidence that lead to optimism about the current strategy. In short, it is an example of *How Institutions Think* (Douglas, 1986).

However, in order to overcome this systemic obstacle, one must first somehow see how they/we are caught up in this "conspiracy" and then seek to change the system itself. Bella's sketches can help us see persistent patterns of behavior that can lead to the "systemic distortion of information." Once we see such patterns of interactions-- systems, contexts, constructions of social realities—we can then ask ourselves if this is the system we want? If not, we must transform it; we must undertake a reconstruction of our social realities.

Ironically, the so-called "conspiracy of optimism" blinds us to the realities of radical reframings. Current realities will themselves ensure that we are unable to address such problematic current realities. In short, we thereby sell ourselves short, ignoring chemical engineer Ilya Prigogine's observation that "the possible is richer than the real" (Prigogine, 1997, p. 72). For unless we address the "conspiracy of optimism" as a systemic phenomenon, we will continue within our current contexts; we will see the only two responses to our environmental problems as naïve optimism or doom-and-gloom pessimism. If, however, we seriously consider radically reframing the current context, then we can at least consider new possibilities, new realities. And in such endeavors, we shall be engaging in 'realistic hope.'

CHAPTER 7

Realistic Hope

In summary, Lackey et al. challenged the authors of *Salmon 2100* to avoid 'delusional optimism' by responding to the four core policy drivers and the question, "What is it *really* going to take to have wild salmon populations in significant, sustainable numbers through 2100?" (Lackey et al., 2006, p. 3). They wanted the book "to play a challenging role that is some mixture of court jester, Greek chorus, cage rattler, and straw man to decision makers, elected and appointed officials, and others who have various mandates and directives to address the decline of wild salmon runs in the Pacific Northwest and California" (Lackey et al., 2006, p. x). To what extent have Lackey et al achieved these goals? This remains to be seen. Bella claims that we need to radically reframe the problem to have realistic hope.

According to Sally Duncan co-author and co-editor of *Salmon 2100* (and Policy Research Program Manager at the Institute for

Natural Resources at OSU), the wild Pacific salmon problems are similar to other larger environmental problems, and many of the same phenomena that one may experience with the salmon problems will be encountered in other environmental problems. Indeed, the wild Pacific salmon crisis is a 'microcosm of the macrocosm' or, as Bella puts it, a "quiz for a larger test." There may be times when one has to be ready in order to transcend and transform the entire system. Simple 'blame' patterns do not solve or dissolve the problems, so we need alternatives to blame. Radical reframing provides such an alternative.

As Cornel West states in *Hope on a Tightrope* (2008), "Shakespeare says in *King Lear*, 'Ripeness is all,' and in *Hamlet*, 'Readiness is all'." According to West, "you become ripe and ready by preparing through tremendous discipline" (West, 2008, p. 126). Because *kairos* times can emerge unpredictably, one must be ready when the time is ripe. Shakespearean scholar James Calderwood concurs with West in "Hamlet's Readiness." Calderwood states,

> Ripeness is a condition at which fruits and flowers
> arrive by nature, but readiness is an achievement;
> the one happens, the other is earned. Even the man
> who is 'ripe' in wisdom seems to simply by some
> inward principle of development to have grown
> wise. But one cannot be 'ready' without having
> readied oneself,... (Calderwood, 1984, p. 267).

For what shall we be ready? That depends on the person and the contexts they are in. But if we give up hope and if we conform to old frames that are no longer effective, then we will be defeated by the 'hopeless dilemma.' Therefore, if we want to live and do something effective, we must radically reframe our problems with realistic hope when the time is ripe. Part of becoming ready is knowing where one is. Bella's context sketches help us to orient ourselves in order to know what we need to change. Once we know where we are, it is easier to transcend and transform the contexts that we do not wish to leave as our legacy. That being said, transformation efforts are not easy as John Kotter, world renowned expert on leadership at the Harvard Business School, points out in "Leading Change: Why Transformation Efforts Fail" (Kotter, 2008). Over the past decade, Kotter has watched more than 100 companies try to remake themselves into significantly better competitors. They included large organizations (Ford) and small ones (Landmark Communications), companies based in the United States (General Motors) and elsewhere (British Airways), corporations that were on their knees (Eastern Airlines), and companies that were earning good money (Bristol-Myers Squibb). These transformation efforts have gone under many banners: total quality management, reengineering, right sizing, restructuring, cultural change, and turnaround. "But, in almost every case, the basic goal has been the same: to make fundamental changes in how business is conducted in order to help cope with a new, more challenging market environment" (Kotter, 2008, p. 1). Through his observations Kotter has learned many lessons:

> The most general lesson to be learned from the
> more successful cases is that the change process
> goes through a series of phases that, in total, usually

require a considerable length of time. Skipping
steps creates only the illusion of speed and never
produces a satisfying result. A second very general
lesson is that critical mistakes in any of the phases
can have a devastating impact, slowing momentum
and negating hard-won gains. Perhaps because we
have relatively little experience in renewing
organizations, even very capable people often make
at least one big error (Kotter, 2009, p. 1).

Based on these lessons, Kotter recommended "Eight Steps to
Transforming Your Organization" (Kotter, 2008, p. 2). In addition
to these steps, we can do what we can do every day, we can
prepare ourselves for a day—*kairos* time or a 'tipping point'—when
our presence will perhaps allow us to do more. Thus, if we have
realistic hope, we will be more ready than if depression and
despair have conquered our will to do anything. If we are "down
and out" then we will not be present with "courage to act" at the
opportune moment even when "doubt is warranted."

In Michael Jackson's movie trailer for "*This Is It,*" Jackson
stated, "this is the moment, this is it; it's an adventure, it's a great
adventure; we wanna' take them places that they've never been
before; I wanna' show them talent like they've never seen before."
Jackson went on later to say, "That is why I write these kinds of
songs. It gives some sense of awareness and awakening and hope
to people." Then he transitions into the song "Man in the Mirror:"

Change, I'm starting with the man in the mirror, oh
yeah, I'm asking him to change his ways. Change,

144

and no message could have been any clearer. If you wanna make the world a better place, take a look at yourself, and then make a change (Jackson, 2009).

My conclusion is that by radically reframing our environmental problems, radical new choices emerge that empower us to transcend the hopeless dilemma and find realistic hope. That hope allows us to be ready when the time is ripe to make a change. Although this book has focused on reframing environmental problems, my frame analysis method and the notion of radical reframing apply to many other kinds of problems as well. For example, we can search for radical reframes in education, economics, health care, military strategies, and business. When the dominant frame is not leading to solutions, then we should search for a radical reframe. Radically reframing problems prepares us to be ready to seize the moment when those ripe *kairos* times appear. Thus, we can have realistic hope and transcend and transform our contexts in ways that leave a more honorable legacy for future generations.

APPENDIX A
Reeves Interview

The following is a summary of the interview I had with Gordon Reeves. I wrote up a summary of our interview and asked Reeves to read it. He did read it, and approved of it as an accurate portrayal of what he said in the interview. The reason I have put this interview here, is that Reeves was asked to participate in *Salmon 2100*, but he refused. So, I wanted to know why. I wanted to know if Lackey et al.'s framing had anything to do with Reeves refusal to participate.

When Bob Lackey asked Gordie Reeves to participate in *Salmon 2100*, Gordie asked him if it was going to be more of his doom-and-gloom perspective. When he responded with silence, Gordie said that he did not want to participate. Gordie stated that Bob's pessimistic framing of the future of wild Pacific salmon in the Pacific Northwest is more based on opinion and interpretation of trends than scientific facts supported by analysis.

In effect, Bob is advocating a policy *prescription*, which is exactly what he says scientists should not do. Bob claims he is like "Joe Friday," just giving the facts. But Reeves states that science is not objective. Every scientist brings biases, and a good scientist will know his or her own biases, make hypotheses about the world, and test them. If the results indicate a different view of reality than the view previously held by the scientist, then the scientist must be able to change his or her hypotheses and views of the world. Bob has brought a particularly pessimistic view and interpretation of the wild Pacific salmon problem to the discussion. But, according to Reeves, this is based more on Bob's opinion than on scientific analysis. Moreover, Bob has "annointed himself as the expert," yet much of his scientific background concerns non-salmonid species.

When I presented Reeves with my maps regarding "realistic hope" and "radical reframing" he agreed that there is a "conspiracy of optimism" and a "conspiracy of pessimism." He also agreed that in response to our environmental problems, these two conspiracies are not the only response. It is a false dichotomy to assume that we must succumb to either the conspiracy of optimism or that of pessimism. When I showed Reeves the notions of radical reframing, where we accept the nonlinear, emergent, and unpredictable aspects of reality, along with 'we talking about us' not just assigning blame to 'them,' he agreed that this is critical.

For example, if he and others had succumbed to doom-and-gloom prior to the unpredictable *kairos* time (Greek notion of the time of opportunity) of the congressional hearings for the Northwest Forest Plan in 1991 (Northwest Forest Plan was adopted in 1994), then they would not have been ready to present the salmon ecosystem science of the time. They were ready, however, and "The Gang of Four" (Gordon, Franklin, Thomas, and Johnson) offered various plans at a congressional hearing on

147

Capitol Hill in Washington. Reeves was an advisor to the "Gang of Four" regarding salmon science. According to Joseph Cone's book *A Common Fate: Endangered Salmon and the People of the Pacific Northwest* (1995), the highlight of the hearing for Reeves was when John Gordon read the congressmen the panel's conclusions: "The current forest plans do not provide a high level of assurance—that is low risk—for maintaining habitat for old-growth-dependent species," Gordon said (Cone, 1995, p. 178).

In their alternatives, Reeves and Sedell were unequivocal about how to save the "at risk" salmon. "Changes in management of federal forests can directly affect the habitat and recovery of those stocks," they wrote (Cone, 1995, p. 180). What is most important about this story is that this opportune moment to affect the ecosystem in which salmon live presented itself unpredictably. Thus, the scientists had to be ready at a minute's notice. Whereas if they had already given up on the salmon struggle, succumbed to doom-and-gloom, then they would not have been ready and present at the Congressional hearing and the *kairos* time opportunity would most likely have been lost. They were ready, however, and now there are 300 foot buffers around the salmon streams, and this has been good for the salmon.

In addition to being ready for the *kairos* times, Reeves made another intriguing comment in our interview about salmon. "They are like weeds," stated Reeves, referring to salmon. "They are tenacious organisms that can endure great disturbances." To give up on them is to deny this fact. If one continues to work with 'realistic hope' and seizes the moment when an opportunity for change arises unpredictably, then one will not have given in to either the 'conspiracy of optimism' or the 'conspiracy of pessimism.'

BIBLIOGRAPHY

Ackoff, R. L. (1999). *Ackoff's best: His classic writings on management* (1 ed.). New York, NY: Wiley.

Ackoff, R. L., Addison, H. J., & Magidson, J. (2006). *Idealized design: How to dissolve tomorrow's crisis...today* (1 ed.). Upper Saddle River, NJ: Wharton School Publishing.

Adams, M., Blumenfeld, W. J., Castaneda, C. R., Hackman, H. W., & Zuniga, X. (2000). *Readings for diversity and social justice: An anthology on racism, sexism, anti-semitism, heterosexism, classism, and ableism* (1 ed.). New York: Routledge.

Armstrong, K. (2001). *The battle for god* (1st Ballantine Books ed.). Chicago: Ballantine Books.

Armstrong, K. (2005). *A short history of myth*. New York: Canongate.

Bardwell, L.V. (1991). Problem-framing: A perspective on environmental problem-solving. *Environmental Management*. 15(5), 603-612.

Beeby-Lonsdale, A. (1996). *Teaching translation from Spanish to English: Worlds beyond words (Didactics of translation).* Ottawa: University Of Ottawa Press.

Bella, D.A. (1996). The pressures of organizations and the responsibilities of university professors. *BioScience,* 46(10), 772-778.

Bella, D.A. (2006). Emergence and evil. *E:CO.* 8(2), 102- 115.

Bella, D.A. (2006). Legacy. in Lackey, R., Lach, D., and Duncan, S., eds. 2006. *Salmon 2100: The future of wild pacific salmon.* Bethesda, Maryland: American Fisheries Society.

Bella, D.A. (2006) *Reframing and legacy: Lessons from Salmon 2100.* Unpublished manuscript.

Benford, R.D. (1997). An insider's critique of the social movement framing perspective. *Sociological Inquiry,* 67 (4), 409-430.

Benford, R., & Snow, D. (2000). Framing processes and social movements: An overview and assessment. *Annual Review of Sociology, 26,* 11-39.

Brewer, J. and Lakoff, G. (2007) The science behind framing. Retrieved January 1, 2009, from http://www.sciencemag.org/cgi/eletters/316/5821/56#9909

Brown, B.C. Integral sustainability 101. Taken from Internet 3/22/07: http://www.integralinstitute.org

Brulle, R. J., & Jenkins, J. C. (2006). Spinning our way to sustainability? *Organization Environment, 19*(1), 82-87. doi: 10.1177/1086026605285587.

Campbell, J. (2008). *The hero with a thousand faces* (2nd ed.). Novato, CA: New World Library.

Carson, R. (2002). *Silent spring* (Anv ed.). New York: Mariner Books.

Capra, F. (1996). *The web of life: A new scientific understanding of living systems.* New York: Anchor Books.

Chantrell, G. (2004). *The Oxford dictionary of word histories.* New York: Oxford University Press, USA.

Chong, D., & Druckman, J. (2007). Framing theory. *Annual Review of Political Science, 10*, 103-126.

Christenden, J. (2004) *What is the meaning of "integral"?* Taken from Internet 4/21/07: http://www.integralinstitute.org

Coase, R.H. (1960). The problem of social cost. *Journal of Law and Economics.* 3, 1-44.

Cone, J. (1995). A common fate: Endangered salmon and the people of the Pacific Northwest. New York: Henry Holt and Company, Inc.

Conner, R.D. and Dovers, S.R. (2002) *Institutional change and learning for sustainable development*. Working paper 2002/1. (Canberra: Center for Resource and Environmental Studies, Australian National University, 2002): www.cres.anu.edu.au/outputs/.

Corning, P. (2002). The re-emergence of "emergence": A venerable concept in   search of a theory. *Complexity*, 7(6), 18-30.

Crutchfield, J., Farmer, D., Packard, N., and Shaw, R. (1986). Chaos. *Scientific American* 255 (December), 46-57.

Czech, B. & Daly, H. (2004) In my opinion: The steady state economy -- what it is, entails, and connotes. *Wildlife Society Bulletin*. 32(2), 598–605.

Daly, H. E. (1978). *Steady-state economics: The economics of biophysical equilibrium and moral growth*. New York: W H Freeman & Co.

D'Angelo, Paul. (2002). News framing as a multi-paradigmatic research program: A response to Entman. *Journal of Communication* 52 (4), 870-88.

d'Anjou, L. (1996). *Social movements and cultural change: The first abolition campaign revisited*. Hawthorne, NY: Aldine de Gruyter.

d'Anjou, L. and van Male, J. (1998). Between the old and the new: Social movements and cultural change. *Mobilization* 3 (2), 207-226.

de Vreese, C.H. (2002). *Framing Europe: Television news and European integration.* Amsterdam: Aksant.

Davis, Murray S. (1975). Review of frame Analysis: An essay on the organization of experience by Erving Goffman. *Contemporary Sociology* 4(6), 509-603.

Diani, M. (1996). Linking mobilization frames and political opportunities: Insights from regional populism in Italy. *American Sociological Review* 61(6), 1053-1069.

DeHaan, R. (2005). The impending revolution in undergraduate science education. *Journal of Science Education and Technology, 14*(2), 253-269.

Development, W. B. C. O. S., & Institute, W. R. (2006). *Greenhouse gas protocol: The ghg protocol for project accounting.* World Resources Inst.

Diamond, J. (2005). *Collapse: How societies choose to fail or succeed.* Penguin (Non-Classics).

DiMaggio, Paul and Walter W. Powell. (1983). The iron cage revisited: Institutional isomorphism and collective rationality in organizational fields. *American Sociological Review* 48 (2), 147-60.

Douglas, M. (1986). *How institutions think*. New York: Syracuse University Press.

Dovers. S.R. (1997). Sustainability; demands on policy. *Journal of Public Policy* 16, 303-318.

Dreyfus, H. (1980). Holism and hermeneutics. *Review of Metaphysics,* 34 (September 1980), 3-23.

Duncan, D., & Burns, K. (2009). *The national parks: America's best idea*. New York: Alfred Knopf.

Durham, Frank D. (2001). Breaching powerful boundaries: A postmodern critique of framing. In *Framing public life: Perspectives on media and our understanding of the social world*, Reese, S., Gandy, O., & Grant, A. Mahwah, NJ: Lawrence Erlbaum Associates.

Eder, K. (1996). Social movement organizations as democratic challenge to institutional politics? And what does this have to do with political citizenship. *Conference on Social and Political Citizenship in a World of Migration* (San Domenico di Fiesole, Italy. Firenze, Italy: European University Institute.

Editors. (2005, September). Science at the crossroads. *Scientific American*, 293(3).

Ehrlich, P. (2009, September 4). MAHB sustainability initiative. Retrieved January 4, 2010, from http://mahbsustainability.wordpress.com/2009/09/

Eisenhower, D. D. (1961). Farewell radio and television address to the American people. In *Public papers of the presidents,* 1035–1040. U.S. Government Printing Office, Washington, D.C.

Entman, R. M. (1993). Framing: Toward clarification of a fractured paradigm. *Journal of Communication,* 43(4), 51-58.

Fairweather, J., & Beach, A. (2002). Variations in faculty work at research universities: Implications for state and institutional policy. *The Review of Higher Education,* 26(1), 97-115.

Fisher, Kim. (1997). Locating frames in the discursive universe. *Sociological Research Online* 2(3), Retrieved January 14, 2010, from http://www.socresonline.org.uk/2/3/4.html.

Freire, P. (1970). *Pedagogy of the oppressed* (15th ed.). New York: Seabury Press.

Friedman, T. L. (2008). *Hot, flat, and crowded: Why we need a green revolution--and how it can renew america* (1st ed.). Farrar, Straus and Giroux.

Ford, J. 1986. Chaos: solving the unsolvable, predicting the unpredictable! In Barnsley, M. and Demko, S., eds., *Chaotic dynamics and fractals,* 1-52. Orlando: Academic Press.

Gamson, W. A. (1975). "Review of frame analysis" by Erving Goffman. *Contemporary Sociology,* 4, 603-607.

Gamson, W.A. (1988). Political Discourse and Collective Action. *International Social Movement Research* 1, 219-246.

Gamson, W.A., & Modigliani, A. (1989). Media discourse and public opinion on nuclear power: A constructionist approach. *American Journal of Sociology* 95(1), 1-37.

Gamson, W.A. (1992). The social psychology of collective action. Pp. 53-76 in: *Frontiers in social movement theory*, edited by A.D. Morris and C.M. Mueller. New Haven, CT: Yale University Press.

Gamson, W.A., Croteau, D., Hoynes, W., & Sasson, T. (1992). Media images and the social construction of reality. *Annual Review of Sociology* 18,373-93.

Gitlin, T. (1980). *The whole world Is watching: Mass media in the making and unmaking of the new left*. Berkeley, CA: University of California Press.

Gitlin, T. (1994). From universality to difference: Notes on the fragmentation of the idea of the left, Pp. 150-174 in: *Social theory and the politics of identity*, edited by C. Calhoun. Cambridge, MA: Blackwell.

Gleick, J. (1987). *Chaos: Making a new science*. New York: Viking Penguin.

Goffman, Erving. (1974). *Frame analysis: An essay on the organization of experience*. New York: Harper & Row.

Goldstein, J. (1999). Emergence as a construct: History and issues. *Emergence: Complexity and Organization, 1*(1), 49-72.

Gould, S. J. (1990). *Wonderful Life: The burgess Shale and the nature of history.* W. W. Norton.

Greenfeld, L. (1999). Is nation unavoidable? Is nation unavoidable today?' Pp. 37-53 in: *Nation and national Identity: The European experience in perspective,* edited by Kriesi, H., Armigeon, K., Siegrist, H., & Wimmer, A. Zürich, Switzerland: Rüegger.

Gunderson, L.H. & Holling, C.S. (2002). *Panarchy.* Island Press Washington D.C.

Hacking, I. (1999). *The social construction of what?* Cambridge, MA: Harvard University Press.

Hallahan, Kirk. (1999). Seven models of framing: Implication for public relations, *Journal of Public Relations Research* 11(3), 205-42.

Harris, G. (2007). *Seeking sustainability in an age of complexity* (1st ed.). Cambridge University Press.

Hawken, P., Lovins, A. B., & Lovins, L. (2000). *Natural Capitalism* (New ed.). Los Angeles: Earthscan Publications Ltd.

Heim, M.H. A life in translation. Lecture by Michael Henry Heim from UCLA at OSU 11/6/09.

Hirt, P. W. (1994). *A Conspiracy of optimism: Management of the national forests since world war two*. University of Nebraska Press: Lincoln.

Holling, C.S. (2007). *Resilience science*. Retrieved March 22, 2007, from http://rs.resalliance.org/

Holland; E. M., Pleasant; A., Quatrano, S; Gerst; R., Nisbet, M. C., & Mooney, C. (2007). The risks and advantages of framing science. *Science, 317*(31 August 2007), 1168-1170.

Hurst, D. K. (2002). *Crisis & renewal: Meeting the challenge of organizational change*. Harvard Business School Press.

Iyengar, Shanto. (1991). *Is anyone responsible? How television frames political issues*. Chicago, IL: University of Chicago Press.

Jackson, M. (2009). *This is it*. Retrieved November 6, 2009, from http://www.thisisit-movie.com/

Janis, I. (1972). *Victims of groupthink: A psychological study of foreign-policy   decisions and fiascoes*. Boston: Houghton Mifflin Company.

Kahneman, D., & Tversky, A. (1979). Prospect Theory: An analysis of decision under risk. *Econometrica 47*, 263-291.

Kauffman, S. (1995). *At home in the universe: The search for laws of self-organization and complexity*. Oxford Univ. Press.

Kauffman, S. (2008). *Reinventing the sacred: A new view of science, reason, and religion.* Basic Books.

Kellert, S. (1993). *In the wake of chaos.* Chicago: The University of Chicago Press.

Kimmel, M.S. (1993). Sexual balkanization: Gender and sexuality as the new ethnicities, *Social Research* 60, 571-587.

Kilgarriff, A. (2005). Language is never, ever, ever random. *Corpus Linguistics and Linguistic Theory, 1(2), 263-276.*

Koenig, T. (n.d.). Reframing frame analysis: systematizing the empirical identification of frames using qualitative data analysis software. Retrieved January 2, 2010, from http://www.allacademic.com//meta/p_mla_apa_rese arch_citation/1/1/0/3/1/pages110319/p110319-2.php

Kohler, R. E. (1991). *Partners in science: Foundations and natural scientists, 1900-1945* (1 ed.). Chicago: University Of Chicago Press.

König, T. (2009) Frame analysis: A primer. Retrieved March 11, 2009, from: http://www.restore.ac.uk/lboro/resources/links/frame s_primer.php

Kotter, J. (1995). Leading change: Why transformation efforts fail. *Harvard Business Review, March-April,* 1-10.

Kuhn, T. S. (1996). *The structure of scientific revolutions* (1st ed.). University Of Chicago Press.

Lackey, R. (2006). Axioms of ecological policy. *Fisheries.* 31(6), 286-290.

Lackey, R. (2007). Science, scientists, and policy advocacy. *Conservation Biology.* 21(1), 12-17.

Lackey, R. (2009) Challenges to sustaining diadromous fishes through 2100: Lessons learned from western North America. *American Fisheries Society Symposium* 69, 609-617.

Lackey, R., Lach, D., and Duncan, S., eds. (2006). *Salmon 2100: The future of wild Pacific salmon.* Bethesda, Maryland: American Fisheries Society.

Lakoff, G. (2004). *Don't think of an elephant!: Know your values and frame the debate.* White River Junction, Vermont: Chelsea Green.

Lazarsfeld, P. (1950). The logical and mathematical foundations of latent structure analysis. in Measurement and prediction. Princeton, N.J.: Princeton University Press.

Leakey, R. & Lewin. R.. (1995). *The sixth extinction: Patterns of life and the future of humankind.* Doubleday.

Leiss, W. (1972). *The domination of nature.* New York, G. Braziller.

Lewin, R. (2000). *Complexity: Life at the edge of chaos* (1st ed.).
University Of Chicago Press.

Lewontin, R. (2002) Stephen Jay Gould—what does it mean to be
a radical? *Monthly Review,* November. Retrieved September
27, 2007, from
http://findarticles.com/p/articles/mi_m1132/is_6_54/ai_94
142087

Luntz, F. (2007). *Words that work: It's not what you say, It's
what people hear* (1st ed.). Hyperion.

Maher, T. (2001). Framing: An emerging paradigm or a phase of
agenda setting. In *Framing public life: Perspectives on
media and our understanding of the social world*, edited by
Reese, S., Gandy, O., & Grant, A. Mahwah, NJ: Lawrence
Erlbaum Associates.

May, R. (1976). Simple mathematical models with very
complicated dynamics. *Nature*, 261, 459-467.

McAdam, D. (1996). Culture and social movements. Pp. 36-57 in
New social movements: From ideology to identity, edited
by Laraña, E., Johnston, H., & Gusfield, J. Minneapolis,
MN: University of Minnesota Press.

Meadows, D., Randers, J., & Meadows, D. (2004). *Limits to
growth: The 30-year update.* White River Junction, VT:
Chelsea Green Publishing Company.

Merkel, J. (2003). *Radical simplicity: Small footprints on a finite earth*. Gabriola Island, B.C.: New Society Publishers.

Meyer, D.S. (1999). Tending the vineyard: Cultivating political process research. *Sociological Forum*, 14, 79-92.

Miller, M. (1997). Frame mapping and analysis of news coverage of contentious Issues. *Social Science Computer Review* 15(4), 367-378.

Miller, M., & Riechert, B. (2001). The spiral of opportunity and frame resonance: Mapping the issue cycle in news and public discourse. In *Framing public life: Perspectives on media and our understanding of the social world*, edited by Reese, S., Gandy, O., & Grant, A. Mahwah, NJ: Lawrence Erlbaum Associates.

Miller, T. R., Baird, T.D., Littlefield, C.M., Kofinas, G., Chapin, F., III, & Redman, C.L. (2008). Epistemological pluralism: Reorganizing interdisciplinary research. *Ecology and Society*, 13(2), 46.

Milgram, Stanley. (1973). The perils of obedience. *Harper's Magazine*, December, 62- 77.

MAHB Mission. (2011) *Millennium alliance for humanity and the biosphere*. Retrieved October 14, 2012.

Mission. (n.d.). *Millennium assessment of human behavior.* Retrieved December 29, 2009, from http://mahb.stanford.edu/mission.html./

Nisbet, M. C., & Mooney, C. (2007). Science and society: Framing science. *Science, 316*(5821), 56. doi: 10.1126/science.1142030.

Nordhaus, T. & Shellenberger, M. (2007). *Break through: From the death of environmentalism to the politics of possibility.* Houghton Mifflin Co.

Oberschall, A. (1996). Opportunity and framing in the Eastern European revolts of 1989. Pp. 93-121 in *Comparative Perspectives on Social Movements: Political Opportunities, Mobilizing Structures and Cultural Framings*, edited by McAdam, D., McCarthy, J.D., & Zald, M.N. Cambridge: Cambridge University Press.

Onians, R. (1973). *The origins of European thought.* New York: Arno Press.

Ornstein, R. (1992). *Evolution of consciousness: The origins of the way we think.* Simon & Schuster.

Perrow, C. (2007). *The next catastrophe: Reducing our vulnerabilities to natural, industrial, and terrorist disasters.* Princeton University Press.

Pepper, S. C. (1961). *World hypotheses: A study in evidence.* Berkeley: University of California Press.

Pielke, R. A. (2007). *The honest broker: Making sense of science in policy and politics.* New York: Cambridge University Press.

Price, V., Tewksbury, D., & Powers, E. (1997). Switching trains of thought: The impact of news frames on readers' cognitive responses. *Communication Research, 24*(5), 481-506.

Prigogine, I. (1997). *The end of certainty: Time, chaos, and the new laws of nature.* (1 ed.). New York City: Free Press.

Prigogine, I., and George, C. (1984). *Order out of chaos.* New York: Bantam Books.

Riedy, C. (2005). *The eye of the storm: An integral perspective on sustainable development and the climate change response.* Dissertation. University of Technology, Sydney, Australia.

Scheufele, Dietram A. (1999). Framing as a theory of media effects. *Journal of Communication, 49*(4),103-22.

Searle, J.R. (1995). *The Construction of Social Reality.* New York: The Free Press.

Semetko, H.A., & Valkenburg, P.M. (2000). Framing European politics: A content analysis of press and television news. *Journal of Communication, 50*(2), 93- 109.

Simon, J. L. (1983). *The ultimate resource.* Princeton: Univ Pr.

Smith, W. (1997). *Modern culture from a comparative perspective*. Albany, New York: State University of New York Press.

Snow, D. A., & Benford, R. (1988). Ideology, frame resonance and participant mobilization. *International Social Movement Research*, 1, 197-219.

Snow, D. A., Rochford, B., Worden, S., & Benford, R. (1986). Frame alignment processes, micromobilization and movement participation. *American Sociological Review*, 51, 464-481.

Somers, M.R. (1995). Narrating and naturalizing Anglo-American citizenship theory: The place of political culture and the public sphere, *Sociological Theory*, 13, 229-274.

Sterman, J. D. (2000). *Business dynamics: Systems thinking and modeling for a complex world* (p. 993). McGraw Hill Higher Education.

Tankard, J.W., Handerson, L., Sillberman, J., Bliss, K., & Ghanem, S. (1991) Media frames: Approaches to conceptualization and measurement. *Paper presented at the Association for Education in Journalism and Mass Communication*, Boston, MA, August 7-10.

Tankard, James W., Jr. (2001). The Empirical Approach to the Study of Media Framing. In *Framing public life: Perspectives on media and our understanding of the social world*, edited by Reese, S., Gandy, O., & Grant, A. Mahwah, NJ: Lawrence Erlbaum Associates.

Tarrow, S. (1992). Mentalities, political cultures, and collective action frames. Pp. 174-202 in *Frontiers in social movement theory*, edited by Morris, A.D. & C. M. Mueller. New Haven: Yale University Press.

Varela, F. J., Thompson, E. T., & Rosch, E. (1992). *The embodied mind: Cognitive science and human experience* (New edition). The MIT Press.

Weaver, W. (1948). Science and complexity. *American Scientist,* 36, 536-544.

West, C. (2008). *Hope on a tightrope.* Carlsbad, California: SmileyBooks.

West, C. (2007, November 14). Interview in *Rolling stone* 40th anniversary issue. Issue, 1039, 116-117.

West, C. (1993, May 30) Commencement address at Wesleyan University Middletown, Connecticut. Full speech retrieved September 21, 2009, from: http://www.humanity.org/voices/commencements/speeches/index.php?page=west at_wesleyan

West, C. (1993). *Prophetic fragments: Illuminations of the crisis in American religion and culture.* Grand Rapids: Wm. B. Eerdmans Publishing Company.

White, E. (1987). *Kaironomia: On the will-to-invent.* Ithaca and London: Cornell University Press.

Whitehead, K. (2007). *Assessing the feasibility of policy prescriptions in the Salmon 2100 project.* Master of Public Policy Thesis at Oregon State University.

Wilber, K. (2001). *A theory of everything.* Shambhala, Boston.

Wilber, K. (2005). *The integral operating system.* Sounds True, Boulder, CO.

ABOUT THE AUTHOR

David Earl Lane is a fifth generation Oregonian whose family came to the Northwest on the Oregon Trail. He has a Ph.D. in Environmental Sciences, Ed.M., B.S., B.A., and a Certificate in Applied Ethics from Oregon State University. David is also a linguistic enthusiast, and has taught both Spanish and English. He and his family enjoy exploring Oregon's rich diversity while tending their garden and home.